AF347516

Building Physics and Building Energy Systems

Building Physics and Building Energy Systems

Editor

Davide Astiaso Garcia

MDPI • Basel • Beijing • Wuhan • Barcelona • Belgrade • Manchester • Tokyo • Cluj • Tianjin

Editor
Davide Astiaso Garcia
Department of Planning, Design,
and Technology of Architecture,
Sapienza University of Rome
Italy

Editorial Office
MDPI
St. Alban-Anlage 66
4052 Basel, Switzerland

This is a reprint of articles from the Special Issue published online in the open access journal *Applied Sciences* (ISSN 2076-3417) (available at: https://www.mdpi.com/journal/applsci/special_issues/Building_Energy_Systems).

For citation purposes, cite each article independently as indicated on the article page online and as indicated below:

LastName, A.A.; LastName, B.B.; LastName, C.C. Article Title. *Journal Name* **Year**, *Volume Number*, Page Range.

ISBN 978-3-0365-0518-3 (Hbk)
ISBN 978-3-0365-0519-0 (PDF)

Contents

About the Editor

Davide Astiaso Garcia is Assistant Professor of Thermal Sciences, Energy Technology, and Building Physics at Sapienza University of Rome. His full operating skills include renewable energies, smart energy systems, energy efficiency in buildings, environmental impact assessments, pollution risk management, and indoor and outdoor air quality monitoring and mitigation measures. He has participated in more than 10 European (Horizon 20202, Interreg Med, ENPI CBC Med, Erasmus+) and national research projects in the field of renewable energies and smart energy systems. He is contracted by the European Commission Research Executive Agency (REA) as an expert for the evaluation of the project proposal under H2020 calls "FET OPEN (Future Emerging Technologies)". He is author of more than 70 scientific publications mainly concerning energy themes (h-index 21 in Scopus database).

Preface to "Building Physics and Building Energy Systems"

The energy transition is one of the key approaches in the effort to halt climate changes, and it has become even more essential in the light of the recent COVID-19 pandemic. Fostering the energy efficiency and the energy independence of the building sector is a focal aim to move towards a decarbonized society. In this context, building physics and building energy systems are fundamental disciplines based on applied physics applications in civil, architectural, and environmental engineering, including technical themes related to the planning of energy and the environment, diagnostic methods, and mitigating techniques. This Special Issue contains information on experimental studies in environmental thermofluid dynamics, problems in environmental comfort, confined environments, the recording and processing of environmental data, and active and passive strategies for environmental monitoring and air-conditioning systems, dealing with the following research topics: renewable energy sources, building energy analysis, rational use of energy, heat transmission, heating and cooling systems, thermofluid dynamics, smart energy systems, and energy service management in buildings. Nine papers were accepted for publication, namely eight research papers and one review paper.

Davide Astiaso Garcia
Editor

 *applied sciences*

Article

Smart Solutions for Sustainable Cities—The Re-Coding Experience for Harnessing the Potential of Urban Rooftops

Valeria Todeschi [1], Guglielmina Mutani [2,*], Lucia Baima [3], Marianna Nigra [4] and Matteo Robiglio [3]

[1] Future Urban Legacy Lab—FULL, Department of Energy, Politecnico di Torino, 10129 Turin, Italy; valeria.todeschi@polito.it
[2] Responsible Risk Resilience Centre—R3C, Department of Energy, Politecnico di Torino, 10129 Turin, Italy
[3] Future Urban Legacy Lab—FULL, Department of Architecture and Design, Politecnico di Torino, 10129 Turin, Italy; lucia.baima@polito.it (L.B.); matteo.robiglio@polito.it (M.R.)
[4] Future Urban Legacy Lab—FULL, Department of Management and Production Engineering, Politecnico di Torino, 10129 Turin, Italy; marianna.nigra@polito.it
* Correspondence: guglielmina.mutani@polito.it; Tel.: +39-011-090-4528

Received: 16 September 2020; Accepted: 9 October 2020; Published: 13 October 2020

Featured Application: This work presents the analysis of the element of the 'roof' as a methodological approach to assess the renovation opportunities of existing rooftops, as a stimulus to mitigate the urban phenomena of heat island mitigation by focusing on the role of codes, policies, and regulations in cities.

Abstract: Urban rooftops are a potential source of water, energy, and food that contribute to make cities more resilient and sustainable. The use of smart technologies such as solar panels or cool roofs helps to reach energy and climate targets. This work presents a flexible methodology based on the use of geographical information systems that allow evaluating the potential use of roofs in a densely built-up context, estimating the roof areas that can be renovated or used to produce renewable energy. The methodology was applied to the case study of the city of Turin in Italy, a 3D roof model was designed, some scenarios were investigated, and priorities of interventions were established, taking into account the conditions of the urban landscape. The applicability of smart solutions was conducted as a support to the review of the Building Annex Energy Code of Turin, within the project 'Re-Coding', which aimed to update the current building code of the city. In addition, environmental, economic, and social impacts were assessed to identify the more effective energy efficiency measures. In the Turin context, using an insulated green roof, there was energy saving in consumption for heating up to 88 kWh/m^2/year and for cooling of 10 kWh/m^2/year, with a reduction in greenhouse gas emissions of 193 tCO$_{2eq}$/MWh/year and 14 tCO$_{2eq}$/MWh/year, respectively. This approach could be a significant support in the identification and promotion of energy efficiency solutions to exploit also renewable energy resources with low greenhouse gas emissions.

Keywords: energy efficiency; smart rooftop technologies; green roof; solar energy; urban heat island; building codes; energy policies; 3D roof city model; urban landscapes

---

1. Introduction

This paper presents some of the results of the Re-Coding research project, undertaken by the Research Centre Future Urban Legacy Lab (FULL) in collaboration with the Responsible Risk Resilience Centre (R3C) of Politecnico di Torino. This project was carried out in support of the current review of the General Development Plan of the city of Torino, Italy. Such research explored the relationship between codes and city morphology with the aim of expanding the scope of the existing building regulatory system to a wider encompassing environmental system of codes that could support the sustainable development of the city. A number of studies across disciplines are currently looking into the role of planning and coding in the definition of policies and regulations for environmental improvement of our built environment [1–4]. Conversely to the traditional planning approach of zoning, such studies discusses the importance of building codes to trigger effective changes on the urban scale by intervening on punctual aspects such as the environmental quality of building parts. Not only within the academic discourse, but also in practice current tendencies of policies and regulation systems tend to focus on the need of environmental awareness with such approach. This is the case, for example, of the cities of New York, with the OneNYC 2050 and Climate Mobilization Act, and of Marseille, with its regulatory planning and coding system, in which punctual actions are determined with simple rules to improve environmental performances. As [1] explained, the complexity of a regulation system might hinder the immediate understanding of the extent and impacts that such regulations have on the built environment, particularly when the overlapping of well-intentioned regulations generated in different time frames result in out-of-date or ineffective rules. For this reason, the Re-Coding project is aimed at redefining the rules starting by the analysis of 'building elements' as the interface between users (i.e., architects, private owners) and institutions. To this end, about 42 elements (i.e., windows, roof, external walls, and others) across scales have been identified and utilized as navigators to allow the mapping of current regulation systems.

This work presents the analysis of the element of the 'roof' as a methodological approach to assess the renovation opportunities of existing rooftops, as a stimulus to mitigate the urban phenomena of heat island mitigation by focusing on the role of codes, policies, and regulations in cities. The results of this analysis and exploration were used to propose and define modifications in the current regulations. Such modifications were aimed at actively promoting sustainable changes in the urban environment, in particular, by providing data to support the modification of the Building Annex Energy Code, currently in use in the city of Turin.

The Premises of the Research

The roof element was analyzed as a device that relates both environmental issues and the revision of the related regulations that define the relationship between urban morphology and the impacts on its ordinary transformation. Moreover, the roof could be considered as the fifth facade of buildings and its surface can be rethought as a platform for multiple uses, action, and potential transformation effects on the city [5,6].

Within this conceptual framework, the focus of the work was to tease out the opportunities offered in rethinking the roof element not only as a separation device between two environments, internal and external, but as a surface capable of catalyzing multiple functions related to urban living.

The analysis of international case studies (e.g., New York, Paris, Marseille, Melbourne, and Lisbon) started in the first phase of the research through the application of a matrix, which allowed the comparison of the design strategies adopted in relation to the current legislation. The matrix utilized is based on the innovation theory applied to architectural and urban design, defined first by Slaughter [7] and later elaborated by Nigra and Dimitrijevic [8]. Such matrix categorized each case study according to type of change, according to the definitions of incremental change, modular change, architectural change, system change, and radical change. These categories define different natures of change. The incremental change is defined as a small change that does not affect the overall nature of an intervention. The modular change is a change that influences a single independent part of a project. An architectural change is a change that alters the relation between major architectural and compositional elements. The system change is a change that alters the overall system functioning by increasing its performance. The radical change is a change that transforms totally the nature of an existing condition. The following table shows the application of these concepts to a number of case studies analyzed, in relation to the type of intervention done on the roof elements on certain projects.

The matrix above (Table 1) highlights the different opportunities to conceive roofs as a resource in relation to both the building system and the city system. The case studies analyzed allowed us to amplify the image of the functions that the roof element can accommodate: Temporary solutions that rethink the roof as a support for site-specific works or a platform hosting light devices for autonomous functions or that extend and intensify the functional program of the building, creating different relationships with the city up to the roof, rethought as a system-generative platform, which increases the building's performance.

Rereading from this perspective, the idea of roofs and the surfaces made available emerges as an additional layer on the city, an infrastructure of latent potential to be activated through a constant dialogue between the project and the regulations [2,3,9].

Following are the main objectives of this work were:

- Showing an example of scientific investigation in support of the re-coding activity undertaken in conjunction with the Turin Municipality.
- Presenting a methodology able to evaluate the potential and feasibility of rooftop renovation in a built-up urban context.
- Evaluating the impact of smart rooftop solutions (insulated roof, green roof, high-reflectance roof, and energy production from solar energy) assessing energy savings, thermal comfort conditions, greenhouse gas (GHG) emissions, and social, environmental, and economic benefits.
- Identifying innovative building codes as an opportunity to promote rooftops' renovation using smart solutions and technologies.

Table 1. Criteria to assess rooftop renovation feasibility.

Case Study		Type of Change
New York: Roof as layer	Modular	OneNYC 2050 and Climate Mobilization Act pushed for converting the majority of city roofs into green layers
Ch2 di Melbourne: Roof as system	System	The roof of this project is conceived to host technical function to improve the overall energy performance of the building
La Friche, Marseille: Roof as City	Architectural	This project change the use of the roof by conferring it the idea of extending the surface of the public city above a private building
'Quel temps fera-t-il demain', Paris: Roof as fith face	Incremental	This project show a small change in the use of the roof, which is treated as the fifth facade of the building by using its surface as a base for street art
MAAT Museum of Art, Architecture and Technology, Lisbon: Roof as infrastructure	Radical	This project offers the example of a roof that, by extending itself to the city becomes an infrastructure

2. Materials and Methods

The methodology described in this section was applied to a case study of the city of Turin. The city is located in the northwestern part of Italy and has a continental climate and almost 900,000 inhabitants. The aim was to assess the applicability of rooftop renovation strategies in a built-up context at district level, investigating environmental, social, and economic impacts of smart roof solutions. Figure 1 describes in detail materials, methods, and tools used in this.

(1) Geographic Information System (GIS) database: The main input data elaborated with the use of a GIS software and the output of the processing.

(2) GIS tools: The tools used to analyze the buildings' characteristics and the urban environment.

(3) Roof suitability: The criteria used to evaluate the roof suitability according to architectural characteristics, morphological context, building codes, and regulations.

(4)   Roof solutions: The most effective rooftop strategies were identified to improve the livability conditions of the city of Turin, and the impact of smart technologies was investigated.

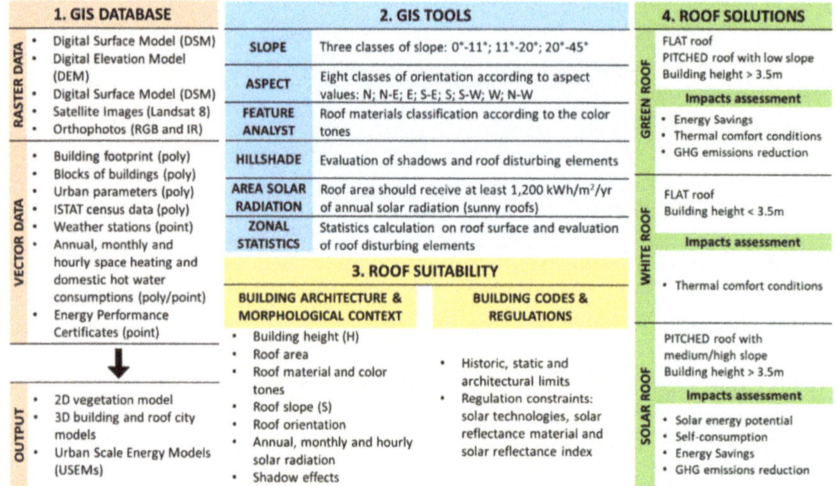

Figure 1. Flowchart of materials, methods, and tools.

2.1. GIS Database: Input Data Collection and Processing

Building upon an ongoing research, a territorial database (DBT) was organized and implemented with the use of a GIS software processing remote sensing images, orthophotos, building characteristics, land cover data, local climate measurements, and energy consumption data. The main data content refers to:

- Elevation models (raster data): The digital terrain model (DTM), with a precision of 10 m, describes the natural terrain. The digital elevation model (DEM), with a precision of 5 m, represents the bare-Earth surface, without natural or built features. The digital surface model (DSM), with a precision of 0.5 m and 5 m, represents the Earth's surface including trees and buildings. These kinds of data were used to assess shadows' effects on buildings and the surrounding's urban context to quantify the solar radiation, taking into account the sun and sky models, and to evaluate the building characteristics such as roof slope and orientation [10].

- Satellite images (raster data) from Landsat 8, the operational land imager (OLI) and the thermal infrared scanner (TIRS) with a precision of 30 m, were used to analyze the land cover types and to calculate the albedo of the outdoor spaces, the presence of vegetation with the use of the normalized difference vegetation index (*NDVI*), and the land-surface temperature (*LST*) [11].

- Orthophotos (raster data), with high spatial resolution of 0.1 m, red-green-blue (RGB) color model, and infrared (IR) spectral bands, were used to identify green areas and evaluate albedo values of outdoor urban spaces and buildings' roofs, as a function of color tones [12,13].

- Municipal technical map of the city (polygonal vector data), updated to 2019, gave information on a building's footprint, type of users, number of buildings, number of floors or building height, period of construction, roof area, gross and net heated volume, net heated surface, and surface-to-volume (*S/V*) ratio. In addition, urban parameters were calculated with buildings' information at blocks of buildings scale [14].

- ISTAT (Italian National Institute of Statistics) census section data (polygonal vector data), updated to 2011, gave information at block-of-building scale on people occupancy, number of inhabitants, number of families and family members, percentage of foreigners, gender, age, income,

- employment rate, socio-economic data (income at 2009), central or autonomous heating systems, and type of fuels.
- Urban parameters (polygonal vector data) at block-of-buildings scale were elaborated using Istat census database and municipal technical maps. The main variables were building density (BD), building height (BH), building coverage ratio (BCR), relative buildings' height (H/H_{avg}), canyon effect (H/W ratio), solar exposition, and main orientation of the streets (MOS) [15].
- Local climate data refers to weather stations' measurements (punctual vector data) located in the city. Available hourly data refer to temperature, relative humidity, vapor pressure, and wind velocity of the outdoor air.
- Space heating and domestic hot water consumption data (punctual and polygonal vector data) were provided by the district heating IREN Company of the city. The annual, monthly, and hourly energy consumptions were processed and georeferenced. These data, used to design and validate urban-scale energy models, refer to three consecutive heating seasons: 2012–13, 2013–14, and 2014–15 [16–18].
- Energy performance certificates (EPCs) (punctual vector data) of the Piedmont Region gave information on residential buildings with 867,131 certificates in about 10 years. These data were used to evaluate the type of energy efficiency action and the impact retrofit interventions for the city of Turin [19].

After the processing of these data, the three main outputs used in this work were a 2D vegetation model, a 3D roof city model, and urban-scale energy models with annual, monthly, and hourly time resolutions.

2.2. GIS Tools: Analysis of Building and Roof Typologies

The analysis of roof typologies and the urban environment was carried out using several tools, explained below.

- Slope tool was used to assess the roof slope of each building using the DSM and the municipal technical map. From the simulation results, the roofs were classified into three categories: (1) Flat roofs with a slope <11°, identified as potential intensive green roofs; (2) pitched roofs with slope ≥11° and <20°, as potential extensive green roofs; and (3) and pitched roofs with slope ≥20° and <45°, as potential solar roofs [13].
- Aspect tool was used to assess the roof orientation using the DSM and the municipal technical map. Eight classes of roof surfaces' orientation were identified according to aspect values (that varied between 0° and 360°). Considering slope values and roof orientation, the pitched roofs were classified into five categories: Gable roofs with North-South (N-S) orientation, gable roofs with East-West (E-W) orientation, hipped/pyramid roofs, shed roof, and half-hipped roof [20].
- Feature Analyst tool was used to analyze roof materials with orthophotos as input data [21] to classify surfaces according to the color tones. In addition, from orthophotos the three bands (red, green, and blue) were analyzed with a GIS tool in order to optimize the classification, identifying dark/black, medium, and light/white roofs' colors.
- Area solar radiation tool was used to quantify the annual and monthly solar radiation values from the DSM. The quota of incident global solar radiation was quantified for each pixel (with a dimension of 0.5 m) and the hours of sunlight were calculated to identify sunny roofs (with three or more hours of sunlight) [22].
- Hillshade tool was used to create a shaded relief from the DSM by considering the illumination source angle and shadows, and in combination with other tools to evaluate roof-disturbing elements.
- Zonal Statistics tool was able to calculate statistics' values of raster data for each roof surface. The roof-disturbing elements, such as dormers and antennas, were identified with the standard deviation using the orthophotos, the annual solar radiation analysis, and the hill–shade analysis.

By overlapping the results of the statistical analysis, the disturbance percentage for each roof was assessed, identifying three classes of disturbance: 15, 25, and 35% [23].

Some outputs of the application of the described methodology are indicated in the following figures. Figure 2a shows the building typologies of a district of Turin with a dimension of 1 km × 1 km. Such classification was made using information on type of users, building height, and the *S/V* ratio values. It is possible to observe that almost 80% of buildings are residential, mainly linear blocks and towers [13]. Figure 2b describes the roof typologies, distinguishing six categories: flat, gable with E-W orientation, gable with N-S orientation, half-hipped, hipped or pyramid, and shed. In this pre-analysis, it was noticed that there is a potential of flat roofs that could be converted into green roofs (Figure 3a), the presence of low buildings with dark surfaces could be converted to light surfaces (high-reflectance roof), reducing the environmental temperature (Figure 2a), and a large quota of residential buildings has an optimal E-W orientation for solar energy production (Figure 3b).

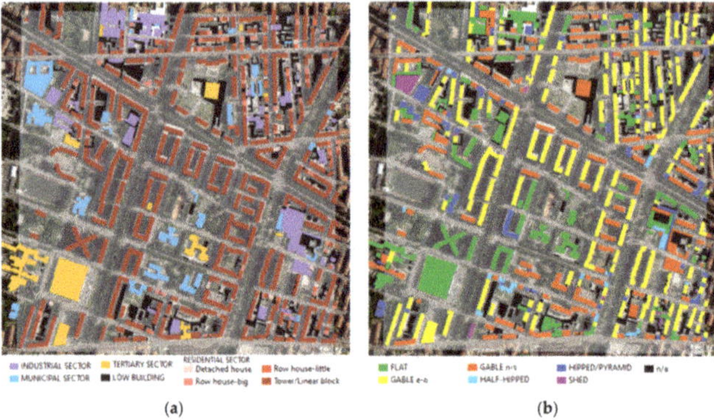

Figure 2. District of Turin with a dimension of 1 km²: (**a**) Identification of building typology using type of users, building height, and surface-to-volume (*S/V*) value; (**b**) identification of roof typologies according to [23].

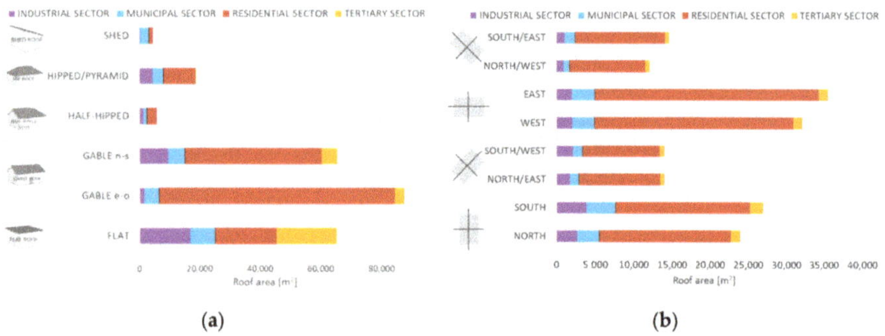

Figure 3. District of Turin with a dimension of 1 km²: (**a**) Roof typology analysis and roof area; (**b**) building orientation analysis and roof area.

2.3. Roof Suitability: Analysis of Criteria to Assess Rooftop Renovations' Feasibility

This section presents the criteria used to evaluate the feasibility of rooftop renovation and to identify the correct rooftop strategy as a function of urban environment. These criteria refer to building architecture, morphological context (Table 2), building codes, and regulations.

The information of buildings' architecture and morphological context were investigated using the DBT presented in Section 2.1. According to Italian Standard (UNI) 11235:2015 and to the literature review [13,24–27], the following criteria were identified to select the potential roofs.

- Building height had to be higher than 3.5 m for green and solar roofs, while for albedo strategies (high-reflectance roof) it had to be less than 3.5 m in order to have the greatest effect on near-surface air temperatures.
- Roof area had to be greater than 100 m^2 for green roofs; for high-reflectance roofs, greater than 20 m^2; and it had to be greater than 50 m^2 for solar roofs.
- Roof material and color tones for green and high-reflectance roofs were excluded; roofs with high reflectance and vegetated roofs, solar roofs, roofs with red tiles and/or disturbing elements, such as dormers and/or antennas, were excluded.
- Roof slope had to be less than 11° (flat roofs) for intensive green roofs and between 11° and 20° for extensive green roofs. There is no limit for high-reflectance roofs and it had to be between 20° and 45° (pitched roofs) for solar roofs.
- Roof orientation with northern exposition was excluded for solar technology, as north-facing rooftops receive less sunlight.
- Solar radiation: Roof area should receive at least 1200 kWh/m^2/year of annual solar radiation for solar technologies. The solar energy potential was investigated, identifying the available rooftop areas and quantifying the total solar radiation on the rooftop.
- Shadow effects: More than 3 h of sunlight for green roofs are necessary to allow the growth of vegetation. Therefore, the shaded roofs (less than 3 h of sunlight) were excluded. In addition, the shadowing effects are important for the selection of the most appropriate plant species for green roofs.

Table 2. Criteria to assess rooftop renovation feasibility.

Criteria	Green Roof	High-Reflectance Roof	Solar Technology
Building height	>3.5 m (heated building)	≤3.5 m (low building)	>3.5 m (heated building)
Roof area	>100 m^2	>20 m^2	>50 m^2
Roof material/color tones	No high-reflectance, vegetated and red-tiled roofs	No high-reflectance, vegetated and red-tiled roofs	No red-tiles roofs No disturbing element
Roof slope	<11° intensive (flat) ≥11° and <20° extensive (pitched)	<8.5° low sloped ≥8.5° steep sloped	≥20° and <45° pitched
Roof orientation	No limit	No limit	No North exposition
Solar radiation	Related to shadow criterion	No limit	≥1200 kWh/m^2/year
Shadow effects	Sunny roofs with more than 3 h of sunlight	No limit	Related to solar radiation criterion

The feasibility of energy efficiency interventions was assessed considering energy and environmental regulations at national and municipal levels. According to the Italian Decree 28/2011, some requirements were considered for the installation of solar energy technologies:

- Production of thermal from solar thermal (ST) collectors' installation: At least 50% of the annual domestic hot water consumption must be covered by the ST production.
- Production of electricity from photovoltaic (PV) panels: The installed electric power, *P*, (in kW) must be greater than or equal to the value calculated with the following equation:

$$P = (1/K) \cdot A \tag{1}$$

where:

P is the installed electric power (kW),
K is a coefficient equal to 50 (m^2/kW) after 1 January 2017, and
A is the footprint area of the building (m^2).

For roofing structures of buildings, verification of the effectiveness, in terms of cost–benefit ratio, was assessed referring to (according to Italian Decree 28/2011):

- Materials with high reflectance of roofs, assuming for the latter a solar reflectance value of not less than 0.65 in the case of flat roofs and 0.30 in the case of pitched roof.
- Passive cooling technologies (e.g., night ventilation and green roofs).

Furthermore, the Solar Reflectance Index (SRI) is used in the main international certification protocols for comparing the coolness of roof surfaces. In Italy some voluntary environmental protocols have been introduced, such as the ITACA (Institute for Innovation and Transparency of Procurement and Environmental Compatibility) protocol, Casaclima Nature certification, and the Green Building Council (GBC) Italia, in which SRI levels for roofs have been specified. In addition, from the enactment of the Italian Decree 11/01/2017, the Ministry for the Environment, Land and Sea has established the "Adoption of minimal environmental criteria (CAM) for the awarding of design services and new construction, renovation and maintenance work on buildings for management of construction sites of the public administration and minimal environmental criteria for the supply of incontinence aids", thus aligning itself with environmental protection strategies adopted at an international level. The section "Reduction of impact on the microclimate and atmospheric pollution" establishes the requirement of materials with a high SRI (Table 3).

Table 3. Italian voluntary protocols and requirements.

Documents	Credits	Application	SRI Threshold Value
LEED 2009 Itaca	1 point	Roofs	At least 75% of the roof surface must consist of material having: $SRI \geq 78$ for low sloped roofs (<8.5°) and $SRI \geq 29$ for steep sloped roofs
GBC HOME	2 points	Roofs	At least 50% of the roof surface must consist of material having: $SRI \geq 82$ for low sloped roofs and $SRI \geq 29$ for steep sloped roofs (>8.5°)
GBC HISTORIC BUILDING	2 points	High-reflectance roofs	
Ministerial Decree 11/01/2017	-	Roofs	$SRI \geq 29$ for roofs with slope greater than 8.5° and $SRI \geq 76$ for roofs with slope less than or equal to 8.5°

The Municipality of Turin regulates the roof elements through a number of rules, as shown in the image below. Current regulations determine rules to design roofs in relation to geometry, structural characteristics, heights, and architectural appearance. Such regulations also define restrictions to design intervention and uses according to functions and zoning of the masterplan, limiting, in particular, changes in the historical center of the city. Moreover, while the Building Annex Energy Code in place calls for environmental awareness by setting compulsory requirements for thermal insulation and derogations to enable the installation of solar and photovoltaic panels, the conversion into green surfaces is only mentioned within the voluntary requirements, leaving the economic burden to the private owners and the limitation of opportunities to out-of-date regulations.

To overcome such limitations, and after the identification of criteria to evaluate rooftops' renovation feasibility, the rooftops' potential was investigated for a district in Turin (IT) and the impact of smart-green technologies was evaluated and quantified using several indicators.

2.4. Impacts of Smart Roof Solutions and Technologies

From the literature review [8,28–37], it emerged that the main roof technologies able to obtain a positive impact on the urban heat island (UHI) mitigation, on the energy consumptions and savings, on the outdoor and indoor thermal comfort conditions, and on social and economic aspects are green and high-reflectance roofs and walls and the energy production from PV panels and ST collectors.

2.4.1. Energy Efficiency Solutions

To evaluate the energy savings after the rooftop renovation, the assessment of heat fluxes through the roof were quantified during the heating and cooling seasons. Different thermo-physical properties of the roofs, indicated in Figure 4, were used according to roof type: Common roof, common insulated roof (red tiles), insulated high-reflectance roof, and insulated green roof.

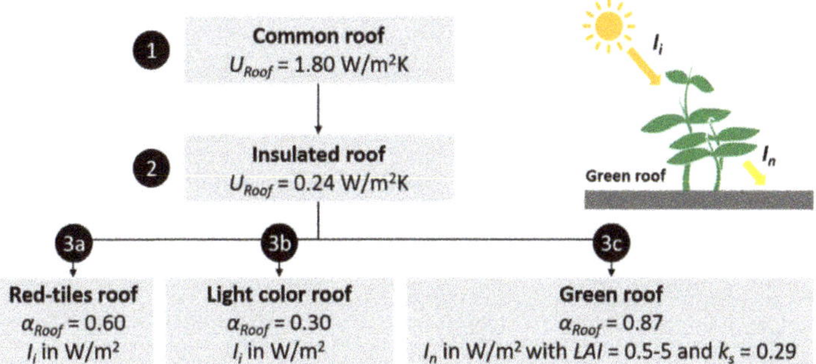

Figure 4. Energy efficiency solutions' scheme.

The roofs have different values of thermal transmittance (U_{Roof}) that depend on the type of insulations. U_{Roof} is taken to equal 1.80 W/m²/K for common roofs and 0.24 W/m²/K for insulated roofs (according to Italian standard) and has different solar absorption coefficients (α_{Roof}) that depend on the roof-covering materials. The α_{Roof} is equal to 0.6 for common roofs with red tiles, 0.3 for light-color roofs, and 0.87 for green roofs. The quota of solar radiation changes according to the presence of vegetation, the incident global solar radiation (I_i), was calculated according to global solar radiation recorded by weather stations, while the quota of incident solar radiation entering a green roof (I_n) depends on the Leaf Area Index (LAI), which is the ratio between the green area and the underneath soil area [38], and on the short-wave extinction coefficient (k_s) [39].

Using green roof technology, the heat flow of solar radiation that enters the system is a net contribution taking into account the solar reflection and green absorption. Equation (2) describes the exponential law developed by Palomo Del Barrio [40] used in this work to assess the effect of green roofs on incident global solar radiation:

$$I_n = I_i^{-k_s \cdot LAI} \tag{2}$$

where:

I_n is the solar irradiance entering the system (W/m²);

I_i is the incident solar irradiance (W/m²);

k_s is the short-wave extinction coefficient (-), which was assumed to equal 0.29 (values proposed for similar vegetation characteristics in [40]); and

LAI is the ratio between the green area and the underneath soil area (-), which was assumed to equal 5 in summer, 3.5 in spring, 3 in autumn, and 0.5 in winter [38,39].

To assess the energy savings of a building, due to the roof component, some simplified assumptions were made: (1) The heat flow rate from internal gains was constant; (2) the heat flow rate dispersed by ventilation was constant; (3) the evapotranspiration of green roofs was not considered; and (4) and the thermal capacity of different roof typologies was equal.

The energy savings for space heating and cooling were quantified calculating the hourly heat flow rates before and after the rooftop retrofit interventions with the following equations [39]:

$$\frac{\Delta Q_H}{A} = U_1 \cdot \left(T_{ai,H} - T_{sa,1}\right) - U_2 \cdot \left(T_{ai,H} - T_{sa,2}\right) \qquad \frac{\Delta Q_C}{A} = U_1 \cdot \left(T_{sa,1} - T_{ai,C}\right) - U_2 \cdot \left(T_{sa,2} - T_{ai,C}\right) \qquad (3)$$

with: with:

$$T_{si} = T_{ai,H} - R_{si} \cdot U \cdot \left(T_{ai,H} - T_{ae} + \alpha \cdot \frac{I}{h_e}\right) \qquad T_{si} = T_{ai,C} + R_{si} \cdot U \cdot \left(T_{ae} + \alpha \cdot \frac{I}{h_e} - T_{ai,C}\right) \qquad (4)$$

where:

ΔQ_H is the energy savings during the heating season (Wh);

ΔQ_C is the energy savings during the cooling season (Wh);

A is the roof area (m^2);

U is the thermal transmittance of the roof (W/m^2/K);

R_{si} is the thermal resistance of the roof (m^{2}K/W);

$T_{ai,H}$ is the internal air temperature during the heating season equal to 20 °C;

$T_{ai,C}$ is the internal air temperature during the cooling season equal to 26 °C;

T_{sa} is the sol–air temperature, which was introduced to take into account not only the external air temperature but also the incident solar irradiation absorbed by the roof (°C);

T_{si} is the internal surface temperature of the roof (°C);

T_{ae} is the external air temperature (°C);

α is the solar absorption of the roof (-);

I_i is the incident solar irradiance (W/m^2), which with green roof was equal to I_n (see Equation (2)); and

h_e is the external thermal adductance (W/m^2/K).

The primary energy savings for space heating and cooling were quantified as the sum of the hourly energy savings during, respectively, the heating and cooling seasons divided by the efficiency of the systems:

$$\frac{\Delta Q_{P,H}}{A} = \frac{\Sigma\, \Delta Q_H}{A} \cdot n_{HS}^{-1} \qquad \frac{\Delta Q_{P,C}}{A} = \frac{\Sigma\, \Delta Q_C}{A} \cdot EER^{-1} \qquad (5)$$

where:

$\Delta Q_{P,H}$ is the primary energy savings during the heating season (Wh);

$\Delta Q_{P,C}$ is the primary energy savings during the cooling season (Wh);

n_H is the average seasonal efficiency of the heating system (in Italy, for residential buildings, this value varies between 0.65 and 0.75 (-)) [17]; and

EER is the average seasonal energy efficiency ratio, which depends on the efficiency of air conditioners (in Italy, for a typical heat pump (air/air) this value is about 3).

Following the energy savings obtained from the retrofit of the rooftop, the GHG emissions' reduction was quantified.

Green Roof Technology

Green roofs alleviate UHI effect through the raising of surface albedo [41] and, so, can reduce the air temperature through evaporation [30] and thermal insulation [42,43]. Therefore, green roofs and walls help in the reduction of energy consumption and guarantee an excellent thermal behavior in both heating and cooling seasons with consequent good thermal comfort conditions, thanks to the high thermal inertia technology [44–47]. Figure 5 shows the heat flows through the roof during the heating and cooling seasons, where Q_{sol} is the quota of solar gains and $Q_{sol,n}$ is the quota of solar gains that enters the system.

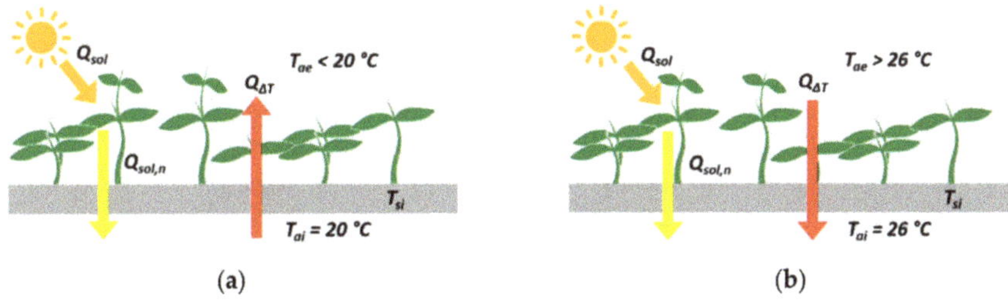

Figure 5. Schematic diagram of green roof during (**a**) heating season and (**b**) cooling season.

The thermal conditions of buildings and urban environments were investigated at urban scale using two parameters: The 'Normalized Difference Vegetation Index' (*NDVI*) and the 'Land Surface Temperature' (*LST*). According to recent studies [31,32], the *LST* and the external air temperature decrease more or less rapidly as the green areas increase, depending also on the type of urban morphology. The local climate conditions were assessed before and after the installation of green technologies.

High-Reflectance Roof Strategy

High reflectance of roofs, identified as albedo strategy, is able to maximize the diffuse reflection of solar radiation, reducing the overheating of buildings and the surrounding urban context and maintaining lower surface temperatures [24,48]. From past studies on UHI mitigation, it has become accepted that a high level of albedo (as white roofs) has the potential to cool cities by 1–3 °C, cooling the lower states/layers of the atmosphere [49–51]. In particular, since in the urban canyon white roof has the greatest effect on air temperatures when used on buildings of 1–2 stories' height [24], in this work low buildings were identified as suitable for this strategy. Moreover, the beneficial effects are greater in a mixed urban morphology context, such as the case of the Turin district analyzed in this work.

As previously mentioned, the *SRI*, used in the main international certification protocols, is a metric for comparing the coolness of roof surfaces. The higher the *SRI*, the cooler the roof will be in the sun [28,52]. For example, a clean black roof usually has an *SRI* of about 0 (with a solar reflectance of 0.05 and an infrared emittance of 0.90), while a clean white roof could have an *SRI* of about 100 (with a solar reflectance of 0.80 and a thermal emittance of 0.90). In general, dark roofs have an *SRI* less than 20 [53].

In this work, the effect of albedo strategy on thermal conditions was investigated calculating the *SRI* and the roof surface temperature (T_s) based on solar reflectance (ρ) and infrared emittance (ε). According to ASTM E1980-11(2019) standard, *SRI* can be defined as:

$$SRI = 100 \cdot \frac{T_b - T_s}{T_b - T_w} \tag{6}$$

with:

$$T_s = 310.04 + 82.49 \cdot \alpha - 2.82 \cdot \sigma - 54.33 \cdot \alpha \cdot \sigma + 21.72 \cdot \alpha \cdot \sigma^2 \tag{7}$$

where:

T_b is the steady-state temperature of a black surface (K) with solar reflectance of 0.05 and infrared emittance of 0.9, under the standard solar and ambient conditions with a solar flux of 1000 Wm^{-2}, ambient air temperature of 310 K, convective coefficient of 12 Wm$^{-2} \cdot$K^{-1} surfaces, and apparent sky temperature of 300 K;

T_w is the steady-state temperature of a white surface (K) with solar reflectance of 0.80 and infrared emittance of 0.9, under standard solar and ambient conditions;

T_s is the temperature of the roof surface (K) under the standard solar and ambient conditions;
α is the solar absorptance of the roof surface (-) equal to $1 - \rho$;
ρ is the solar reflectance of the roof surface (-); and
σ is the Stefan–Boltzmann constant, 5.67×10^{-8} (Wm$^{-2}\cdot$K^{-4}). Table 4 shows typical roofing materials with solar absorption (α), solar reflectance (ρ), and infrared emittance (ε) values used in this work to quantify SRI and T_s before and after roof renovation using the albedo strategy.

In the analyzed district, the values of roofing material properties refer to 'generic black shingle' for dark and black roofs, 'gray Ethylene-Propylene Diene Monomer (EPDM)' for medium roofs, and 'white EPDM' for white and renovated roofs. The values of SRI and T_s were calculated both at building scale and at blocks-of-building scale to evaluate the external conditions.

The main problem of this strategy is that over time the solar reflectance values of high- reflectance roofs decrease due to the accumulation of surface dirt and the degradation of the material by about 0.15 mainly during the first year [54]. The emission, however, does not decrease significantly, and washing the roof surfaces could restore the roof solar reflectance to 70%–100% of the original values [55].

Since most roofs are not washed frequently, it is necessary to evaluate aged values of solar reflectance and infrared emittance values to predict energy savings. If aged values of a roof are unknown, it is possible to estimate the aged solar reflectance ($Aged_\rho$) based on the initial solar reflectance ($Initial_\rho$) by using the following equation:

$$Aged_\rho = 0.7\cdot\left(Initial_\rho - 0.2\right) + 0.2 \tag{8}$$

Table 4. Solar performance of roofing materials [48,56].

Roof Material	α (-)	ρ (-)	ε (-)	T_s (°C)	SRI (-)
Smooth bitumen	0.94	0.06	0.86	83	−0.1
Generic black shingle	0.95	0.05	0.91	82	0.1
Vegetated field	0.90	0.10	0.76	83	−0.2
Grey EPDM	0.77	0.23	0.87	68	0.21
Red clay tile	0.67	0.33	0.90	69	0.36
Red concrete tile	0.82	0.18	0.91	76	0.17
Shasta white shingle	0.74	0.26	0.91	64	0.27
Light gravel	0.66	0.34	0.90	57	0.37
Aluminum	0.39	0.61	0.25	48	0.56
White EPDM	0.31	0.69	0.87	25	0.84
White coating on shingle	0.29	0.71	0.91	23	0.87
White PVC	0.17	0.83	0.92	11	1.04

Referring to LEED (Leadership in Energy and Environmental Design) environmental protocol is also possible to assess mixed nonroof and roof measures, using the following relation as a function of area surfaces (A):

$$\frac{A_{nonroof\ measures}}{0.5} + \frac{A_{high\ reflectance\ roof}}{0.75} + \frac{A_{vegetated\ roof}}{0.75} \geq A_{total\ site} + A_{total\ roof} \tag{9}$$

Solar Energy Technology

Solar thermal (ST) collectors and photovoltaic (PV) panels that exploit renewable energy sources (RES) provide environmental and economic benefits [57–59].

In this work, the solar technologies were assessed considering the potential roofs' area with better solar exposition. In Italy, the most used types of low-temperature ST collectors are *flat glass collectors* with high efficiency and low cost and *vacuum tubes*, which have greater efficiency compared to flat glass collectors, due to the lower dispersions by thermal convection inside the vacuum tubes, but higher cost [60]. On average, a solar thermal system in Italy has a monthly efficiency of 40–85% with flat collectors and 70–86% with vacuum tubes; a collection area of 0.7–1.2 m^2/person for flat collectors and 0.5–0.8 m^2/person for vacuum tubes (considering the production of domestic hot water), with the month of maximum solar radiation being considered for the dimensioning; and a cost of 1000 euro/m^2 for flat collectors and 1200 euro/m^2 for vacuum tubes.

Regarding PV modules, the efficiency of converting solar energy into electricity varies mainly according to the type of technology chosen. The average efficiency values vary from 22% (high efficiency *monocrystalline silicon*) to 4% (*amorphous silicon*). The cost of a PV solar system depends on the installed power, which is around 2000 euro/kWp where kWp is the peak power). The capturing surface depends on the efficiency of the module and ranges from 5.5 m^2/kWp for high efficiency monocrystalline silicon to 11 m^2/kWp for amorphous silicon.

In assessing the efficiency of converting solar energy, it is also necessary to consider the energy losses of all system components, in addition to solar panels; it is estimated to be around 20–25%.

The GIS tool 'Area solar radiation' was used to quantify how much solar radiation each rooftop in the district receives throughout the year. The sun and sky models were elaborated, considering the monthly data of atmosphere transparency (τ) and ratio of diffuse radiation to global radiation (ω) identified from the 'Photovoltaic Geographical Information System PVGIS' of Joint Research Centre (JRC). In particular, considering the period 2013–14, τ was taken to equal 48%, 62%, and 72% in winter, midseason, and summer periods, respectively, and ω was taken to equal to 48%, 45%, and 35%, similarly.

According to European Standard (EN) 12975-2:2006 and Italian Standard (UNI) 11300-4:2016, the two typologies of ST collectors have, on average, respectively, zero-loss efficiencies η_0 of 0.94 and 0.88, linear heat loss coefficients a_1 of 3.34 and 1.57 W/m^2/K, quadratic heat loss coefficients a_2 of 0.02 and 0.01 W/m^2/K^2, and, for the whole system, a performance ratio of 75%. Then, their monthly efficiencies vary from 0.37 to 0.87 for the flat glass collectors and from 0.69 to 0.87 for the vacuum tubes.

The PV modules have an efficiency of 15% (standard efficiency polycrystalline silicon module), and both PV and ST have a system performance around 75%. The hypothesized ST areas were dimensioned in order to not have an overproduction of hot water during summertime.

The monthly energy consumption was simulated for a district of Turin as follow:

- For the residential sector, space heating consumption refers to measured data for the season 2013/2014 [17] and domestic hot water consumption was calculated taking into account that a person needs 50 L of water per day at a temperature of 45 °C (water temperature variation is 30 °C). For the nonresidential sector, space heating and domestic hot water consumption were quantified knowing, for different users, the specific consumption in kWh/m^3 and the heated volume (m^3) [16];
- For the residential sector, electrical consumption refers to the average monthly consumption of 1206 families for the years 2013 and 2014 [61]. For the nonresidential sector, electrical consumption (kWh$_{el}$) was quantified knowing specific annual consumption in kWh$_{el}$/m^3 and the heated volume (m^3) [62].

3. Results

This section describes the main results obtained by applying the methodology presented to a district of the city of Turin, 'Pozzo Strada'. Turin is located in the northwest of Italy, in the Po valley, and it is characterized by a temperate-continental climate, with cold winters and a shorter but hot summer. According to Italian standard UNI 10349:2016, Turin's climate is characterized by 2648 heating degree day (HDD) at 20 °C and 84 cooling degree day (CDD) at 26 °C. The results of this study were presented for a district with a dimension of 1 km^2 with 21,520 inhabitants and more than 1000 buildings.

3.1. Model Application

In the selected district, 1228 buildings were analyzed. Of these buildings, 1097 were classified as potential rooftop renovation opportunities, distinguishing three types of smart solutions: Green roof technology, high-reflectance strategy, and energy production from ST collectors and PV modules.

According to [23], in order to give a priority of interventions, critical areas with the worst air quality conditions were identified as priority areas for the installation of green roof technologies (Figure 6a, red areas) to mitigate the UHI effect. The other areas, with mainly residential buildings, were considered for solar energy production using ST collectors and PV panels. Solar technologies were dimensioned considering residential and nonresidential demand. Figure 6b shows the rooftop classification, distinguishing these three types of smart solutions.

The main characteristics of the buildings selected as potential are indicated in Table 5. It is possible to observe that, thanks to the typical urban mix of Turin, the retrofit measures are well distributed within the district and, moreover, there is a consistent potential. For this reason, it is important to encourage the buildings' renovation – in this case the rooftop renovation—especially in consolidated urban contexts where energy efficiency measures to intervene on buildings are limited.

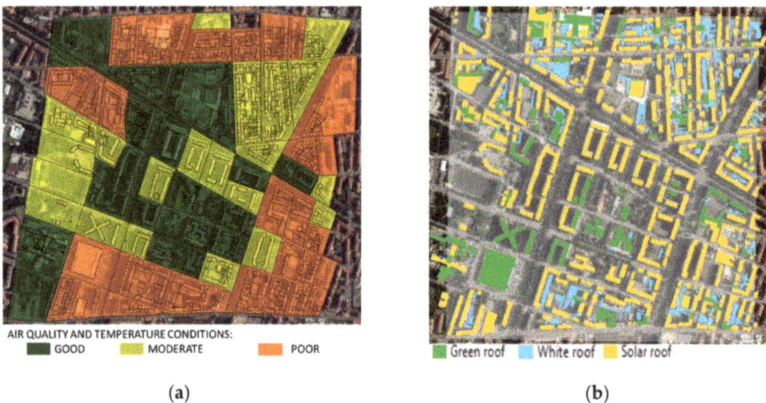

Figure 6. District of Turin with a dimension of 1 km^2: (**a**) Building block classification according to three classes of air quality conditions (green, good; yellow, acceptable, red, bad) [23]; (**b**) analysis of roof potential and feasibility of smart solutions: Green, high-reflectance, and solar roofs.

Table 5. Buildings' characteristics.

Roof Solutions	No Buildings	Height$_{avg}$ (m)	Potential Roof Area (m^2)	Slope$_{avg}$ (°)
Green roof	110	13.6	64,712	0
High reflectance roof	417	3.6	44,956	9
Solar roof	570	19.3	172,749	36

3.2. Smart Roof Solutions' Assessment

This subsection describes the main results obtained from the use of three smart solutions: Green roof technology, high-reflectance roof strategy, and solar energy technology. The aim was to harness the potential of urban rooftops in a district in the city of Turin.

3.2.1. Green Roof Technology

In the district analyzed, 64,712 m^2 of roofs were identified as potential green roofs. Referring to Equation (3), the energy savings for heating and cooling seasons were quantified for a district in

Turin. In this scenario, potential roofs were renovated using green roof technologies. The thermal transmittance with green technologies is equal to 0.24 W/m^2/K (according to Italian Decree 26/6/2015) and the solar absorptance of a green roof surface is 0.87 [48,56]. The energy savings after the installation of green roofs was equal to 5669 MWh/year, which corresponds to 8.4% of space heating consumptions of residential buildings. The energy savings during cooling season was equal to 662 MWh/year. Figure 7 describes the energy savings at block-of-building scale, distinguishing heating and cooling seasons.

Thermal conditions were investigated using some parameters calculated at block-of-building scale from satellite images (Section 2.1.). These parameters are the *NDVI* and the *LST* and allow us to describe the UHI effect and the local-climate characteristics of the urban environment. An analysis at blocks-of-building scale was made, and Figure 8a shows the variation of *LST* before and after the installation of green roof technologies. According to the literature review [34,35], the *LST* and the air temperature tend to decrease more or less rapidly as the green areas increase, depending also on the type of urban morphology. Increasing the green roofs' areas of 64,712 m^2, on average, the *LST* in the district tends to decrease by 1 °C (Figure 8b).

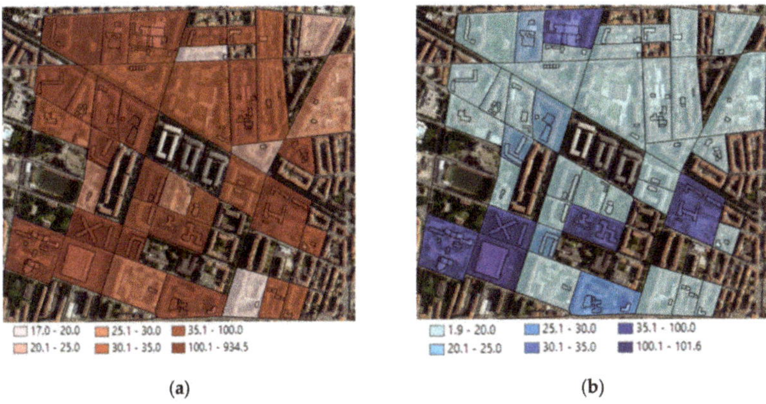

Figure 7. Green roofs' potential assessment at block-of-building scale: (**a**) Heating and (**b**) cooling primary energy savings in MWh/year.

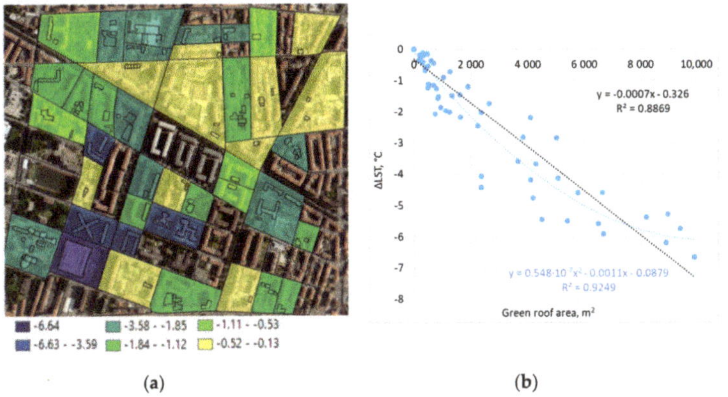

Figure 8. Green roofs' potential assessment at block-of-building scale: (**a**) Thermal condition assessment, Land Surface Temperature (*LST*) variation before and after the installation of green roof technologies; (**b**) correlation between the *LST* variation and the quota of green roof area.

The feasibility of green roof technology was assessed considering requirements the Ministerial Decree 26/06/2015. The roofs' albedo in the district analyzed varied between 0.05 and 0.26 (for a few buildings, mainly industrial, the albedo was around 0.33) [23]. After retrofit measures with green roofs, the roof albedo criterion was respected due to the installation of passive cooling technology.

3.2.2. High-Reflectance Roof Strategy

Starting with 500 low buildings located in the district of Turin, 417 were selected as potential for the renovation of rooftop with white color (high-reflectance roof). Of these 417 potential buildings, which corresponded to an area of almost 45,000 m^2, 313 had a slope less than 8.5° and 104 had a higher slope (on average, had slope of 8.8°, see Table 5). Figure 9 shows the *SRI* values calculated for each block of buildings. The *SRI* values are weighted according to the m^2 of each roof. Figure 9 shows the *SRI* roof values at block-of-building scale, before (Figure 9a) and after (Figure 9b) the use of high-reflectance roof strategy on 417 potential roofs. From the results, it emerged that it is possible to obtain an increase in *SRI* of almost 30 and a reduction of T_s of over 10 °C. Therefore, these indicators could help designers and consumers to choose the proper materials for sustainable buildings and communities.

The feasibility of high-reflectance roof strategy was assessed according to the Italian Decree 11/01/2017 and the environmental protocols. The *SRI* prerequisites (*SRI* > 0.29) were respected.

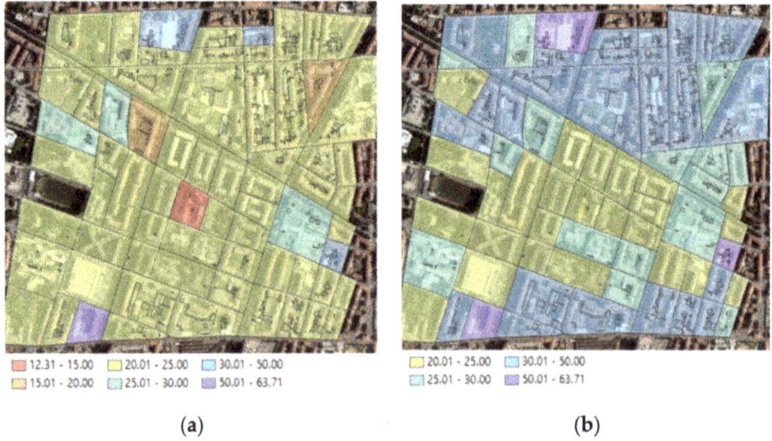

Figure 9. Solar Reflectance Index (*SRI*) (%) values of existing roof at block-of-building scale: (**a**) Before (**b**) and after high-reflectance strategy.

3.2.3. Solar Energy Technology

After the analysis of monthly and annual solar radiation on each rooftop (Figure 10), taking into account 570 heated, pitched buildings and roof surface with annual solar radiation higher than 1200 kWh/m^2/year, the ST collectors and PV modules were dimensioned according to domestic hot water consumption and electrical consumption of residential and nonresidential users.

According to the Italian Decree 28/2011, 50% of domestic hot water consumption of residential sector is covered by ST collectors. The percentage reaches 100% in June, while in the winter months (December and January) ST production is able to cover about 7% of the residential consumption (Figure 11a). In addition, there is a GHG emission reduction of 1958 ton/CO_2/year. The requirements indicate that 50% of consumption must be covered; this dimensioning is appropriate, given that in July hot water can only be used to cover domestic hot water.

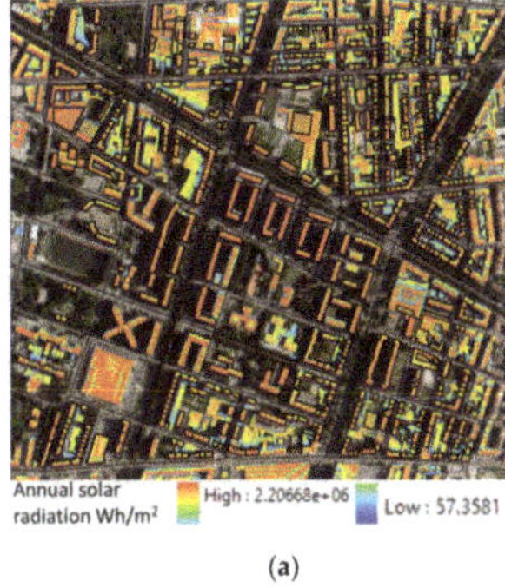

(a) (b)

Figure 10. Solar energy technology assessment: (**a**) Annual solar radiation, solar roofs are identified with a black outline; (**b**) identification of areas (in red) with annual solar radiation <1200 kWh/m^2 (not suitable for solar energy production).

The PV panels can be dimensioned in two ways: By covering 100% of consumption in the month of maximum irradiation or by reaching 100% of annual self-consumption, taking into account that the overproduction in the summer months that is fed into the grid will be consumed in winter. In this work, the PV panels were dimensioned according to the National Decree 28/2011 using the footprint area of the buildings (A), where the installed power is equal to A divided by a K coefficient = 50 (Figure 11b). According to the installed power and the annual utilization hours of use (in the Piedmont region are 1130 h), the electricity produced from PV panels was compared to the electrical consumption. Therefore, knowing that a typical Turin family needs about 2000 kWh$_{el}$/year for electricity supply and, in the district analyzed, the number of families is equal to 10,638 (ISTAT data, 2011), the 13% of the annual residential and nonresidential electrical consumption has been covered with a GHG emission reduction of 1853 ton/CO$_2$/year. Considering only the residential sector, PV production covers the 18% of electricity consumption. In the summer months it covers 38% and in the winter months, 2–3%. Table 6 shows the total roof area, the quota well exposed with no disturbances (15–35%), the quota used for the ST collectors to satisfy the domestic hot water (DHW) consumptions, and the quota for the PV panels, as requested by the standards (1/50 kW/m^2). Using the maximum energy potential that can be produced from PV panels with the left available roof area, it is possible to cover 82% of residential electrical consumption; with the reverse procedure, an optimal value of K of about 11 m^2/kW was calculated.

Table 6. Roof area for solar energy production.

Area	Roof	Well Exposed with No Disturbances (15–35%)	ST for DHW Energy-Use	PV 1/50 kW/m^2	PV max
m^2	172,749	101,048	4717	21,141	96,631
	100%	58%	2.7%	12.2%	55.8%

From this analysis, it emerged that to reach the 50% coverage of domestic hot water, heating, and cooling consumption, it is necessary to use not only solar technologies but also other renewable energy technologies, such as the energy taken from the cold source with heat pumps for heating or PV panels for cooling. In fact, in the city, there are few renewable energy sources available, but there are sources that can be exploited in public spaces, such as the PV panels on shelters, micro-power plants (of which in Turin city there are three) and mini-wind on commercial buildings, considering the acoustic impact.

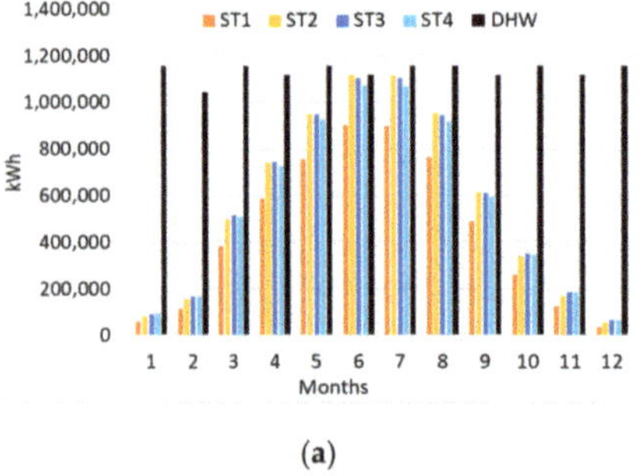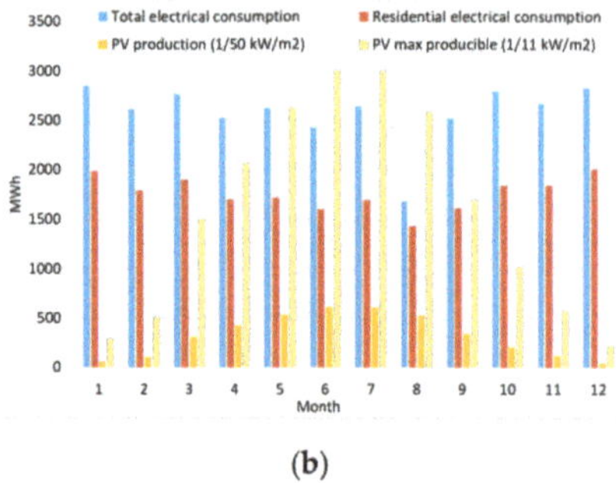

Figure 11. Solar energy technology assessment (for the 2014 year): (**a**) Comparison between domestic hot water (DHW) consumption of residential sector and solar thermal (ST) production considering four collector typologies (collectors' annual average efficiency: ST1 = 0.59, ST2 = 0.77, ST3 = 0.80, and ST = 0.79); (**b**) comparison between electrical consumption, photovoltaic (PV) production with coefficient $K = 50$ m^2/kW (according to the Decree 28/2011), and PV max producible.

3.3. Energy Savings: Heating and Cooling

Green and high-reflectance roofs (cool roofs) have a significant effect in reducing energy consumption during cooling and heating seasons. In accordance with literature review [38,63–67], from this work it emerged that cool roofs are more effective in reducing heat gain in the cooling (C) season from 15 April to 14 October, than heat loss in the heating (H) season from 15 October to 14 April (Figure 12).

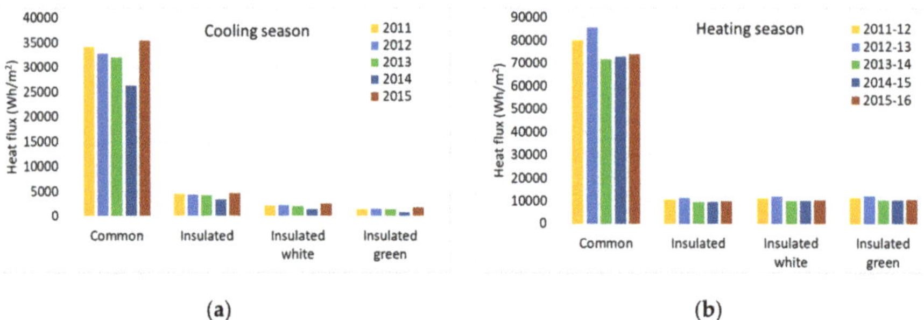

Figure 12. Comparison of heat fluxes (Wh/m^2) between common roof, insulated common roof, insulated high-reflectance roof, and insulated green roof: (**a**) Cooling season; (**b**) heating season.

This analysis was carried out using weather data measurement recorded by Politecnico weather station (WS) for the period from 2011 to 2016. The thermal performance of a refurbished roof was compared to the typical common roof. In particular, three roof solutions were taken into account: (1) Insulated common roof, (2) insulated high-reflectance roof, and (iii) insulated green roof. The heat flux (Q) in the roof was quantified according to Equation (3). Table 7 describes the characteristics of roof solutions and the main energy efficiency results. GHG emissions were quantified using 0.210 tonCO$_2$/MWh for natural gas and 0.46 tonCO$_2$/MWh for electricity [68].

Table 7. Characteristics of roof solutions and energy efficiency results.

Roof Solutions	A (-)	U (W/m²/K)	Q_H (Wh/m²)	Q_C (Wh/m²)	ΔQ_H (Wh/m²)	ΔQ_C (Wh/m²)	GHG_H (tCO₂/MWh)	GHG_C (tCO₂/MWh)
Common	0.60	1.80	76,838	32,135	-	-	1333	319
Common insulated	0.60	0.24	10,245	4285	88,790	9284	178	43
Insulated white	0.30	0.24	10,874	2147	87,951	9996	189	21
Insulated green	0.87	0.24	11,130	1457	87,611	10,226	193	14

Figures 13 and 14 show the results for three consecutive hot days (21–23 July 2015) and cold days (15–17 January 2012). From the comparison of hourly heat fluxes (W/m²) between common roof, insulated common roof, insulated high-reflectance roof, and insulated green roof, it emerged that using an insulated green roof there was less heat gain during the summer season and with an insulated roof there was less heat loss in the winter season.

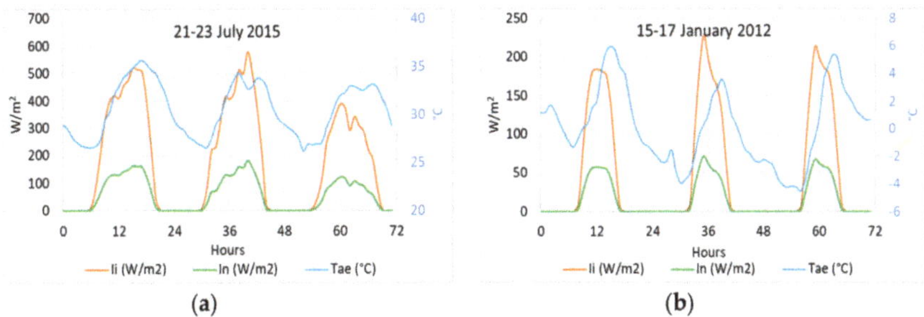

Figure 13. Hourly values of global solar radiation (I_i), solar radiation entering in the system (I_n), and the external air temperature (T_{ae}) for three consecutive days: (**a**) 21–23 July 2015; (**b**) 15–17 January 2012.

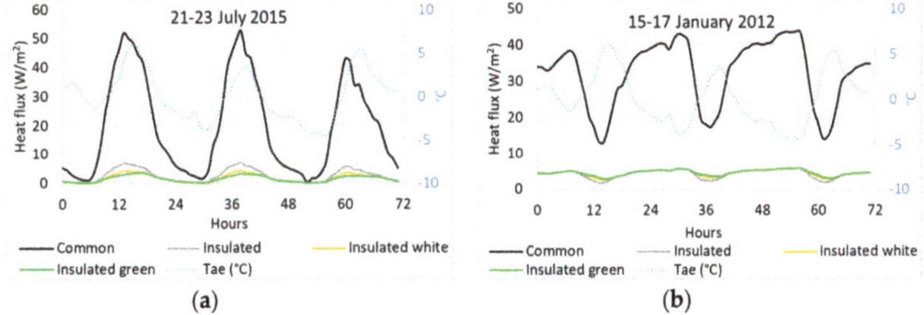

Figure 14. Comparison of hourly heat fluxes (W/m²) between common roof, insulated common roof, insulated high-reflectance roof, and insulated green roof for three consecutive days: (**a**) 21–23 July 2015; (**b**) 15–17 January 2012.

This methodology, used to evaluate the effect of green and cool roofs, will be implemented in future work, adding the effect of evapotranspiration on energy performance of a building.

4. Discussion

The project Re-Coding represents an experience of updating current codes and normative framework to shape the morphology of a sustainable city. This project approach is a multiscale method that, from code design to city scale, can actively trigger sustainable impact by design. The relevance of this work is to be found in a number of aspects that can be generalized and suggested for future processes of normative updates to support sustainable development and environmental resilience, such as in the example shown about the roofs. In particular, a number of lessons can be drawn out from this experience and shared as general knowledge: (1) The institutional vision and wish to review and implement codes and normative framework is crucial to foster environmental sustainability change and resilience [69,70] and (2) the work of multidisciplinary research centers such as *FULL* and R3C allowed us to analyze the understanding of the local normative framework and to update it on scientific-based solutions related to global-based knowledge advancement. Moreover, this work was crucial to increase the interfaces between clients, institutions, and practitioners, through the study of specific building elements, such as the one of the roofs. In this regard, although the methodology was applied to a limited number of neighborhoods in Turin, it was possible to highlight the potential of the roof surfaces of the overall city. This was possible, in particular, by understanding the relation between such potential changes and the current normative framework in use. Despite few rules limiting the use of roofs in the city, concerned mainly with the zoning of the city center, most of the potential uses are actually possible. Figure 15 summarizes the application of the results found after the application of the methodology that was utilized in support of the decision-making process that the Municipality of Turin undertook to revise its local environmental regulation. Moreover, the proposed methodology showed that it is possible to extend actions both to improve the energy performance of the building and environmental performance and to increase the intensity of use of flat surfaces and the related social impact. A latter process could be triggered widely through subjective and individual actions of the owners of the space under the roof that could be the potential users of that space or open these spaces to an urban and collective dimension. As explained in the following sections, many are the environmental, social, and economic benefits by the potential use of roof surfaces.

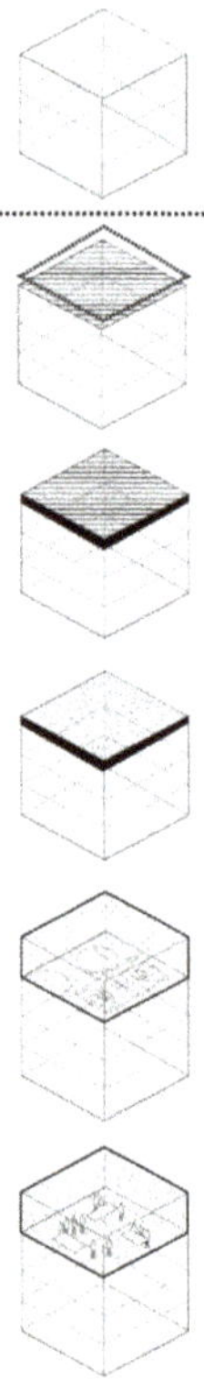

	Strategy	Observation	Potential	Actions
	Color/material change	Admitted but not incentivized	Environmental/social	Creating an incentive
	Performance improvement	Admitted and incentivized	Environmental/social	Incentivized
	Green rooftop	Admitted and incentivized	Environmental/intensity of use / social	Incentivized
	Temporary and light perimeter structures	Allowed but limited (e.g. to respect the height of the fronts)	Intensity of use/social	Exceeding current regulations
	Platform for intensity uses	Admitted	Intensity of use / social	Increase in incentives

Figure 15. Application of the results: Re-Coding activities.

4.1. Impact Assessment of Urban Rooftops' Renovation

The achievement of resilient and sustainable cities depends on several factors, such as the characteristics of territory, the urban morphology, the energy performance of buildings, and the existing laws, constraints, and regulations. There is no one solution to reach this goal, but there are different low-carbon strategies, some of which are presented in this work. In particular, three rooftop renovation strategies are presented, and the effect of the use of these strategies was assessed for the city of Turin at district level.

4.1.1. Environmental, Social, and Economic Benefits

As it will be highlighted in the following subsections, the environmental, social, and economic impacts of rooftop 'green' renovations are widely recognized. Such benefits could be summarized as energy consumption reduction, UHI and air pollution mitigation, water management, sound insulation, and noise reduction, as well as ecological preservation, real estate market positioning improvement, building value increase, and psychological effects on the direct and indirect users, triggering economic and social activities, factors of enjoyment, concentration improvement, crime reduction, and productivity and creativity improvement. [8,36,37,71–73].

Environmental Benefits

At the environmental level, Berardi et al. [71] explained that the benefits of green roofs are multiple, such as: Energy consumption reduction, UHI and air pollution mitigation, water management, sound insulation, and noise reduction, as well as ecological preservation.

In the literature, several studies investigated energy savings and environmental benefits after the use of green roofs, high-reflectance roofs [48,65,73–77], and solar technologies [55–57]. For example, increasing urban albedo can reduce summertime temperatures, resulting in better air quality and savings from reduced air-conditioning costs [78]. Akabari et al. [78] found that converting 100 m^2 of dark roof to white offsets the emission of ~10 tons of $CO_{2,\ equivalent}$. Moreover, 100 m^2 of green roof has a one-time global warming offset potential of 3–4 tons of $CO_{2,\ equivalent}$ [79].

From the results of this work, it emerged that, through the retrofit of the roof, using, for example, green roof, environmental benefits can be obtained, thanks to the reduction of space heating and cooling consumption and, consequently, the reduction of GHG emissions with an improvement of urban thermal conditions. In future work, more detailed analyses will be made for some buildings' blocks located in the same district using the ENVI-met software [80]. This tool, based on a holistic model, is able to simulate local climate conditions in an urban environment and to assess the effects of green technologies on energy savings and thermal comfort conditions.

Economic Benefits

The economic benefits of green roofs are widely discussed in literature over at least a decade [71–73,81–86].

Teotónio et al. [81] explained that different economic evaluations of green roof systems give a high variability of results. Moreover, Teotónio et al. [81] reported that, from a financial perspective, green roofs' investments usually lead to financial losses of 19–50%, given their limited private benefits, whereas, from an economic and socio-environmental perspective, green roofs are mostly identified as interesting investments, leading to gains of 24–40%. Castleton et al. [73] reported that, in terms of whole-life cost analysis, the Net Present Value (*NPV*) of a green roof is 10–14% more expensive than a conventional roof over a 60-year lifetime. Yet, Bianchini et al. [82] explained that a reliable encompassing lifecycle net benefit-cost analysis, or any other calculation systems, should also take into consideration personal costs and benefits, initial construction cost, property value, tax reduction systems in place, storm water retention and avoidance in drainage system, energy reduction (both in terms of cooling and heating), plant longevity benefits, and operational and maintenance costs. Moreover, Sproul et al. [79] demonstrated a considerable variation in the economic value of white, green, and black flat roofs.

Bianchini et al. [82] also stated that:" … Green roofs provide personal and social benefits … there is a low financial risk for installing any green roof type. Additionally, from a personal perspective, the potential profit of an intensive green roof is much higher than its potential losses. Vegetative roofs are a personal investment. However, over the lifecycle of these roofs, both personal and social sectors derive economic benefits. In fact, when social costs and benefits are considered in the NPV estimation, the profitability of the investment is higher. Installing green roofs would be an even more attractive business, if social benefits were partially transferred to investors. The governments should promote green roof construction by reducing insurance premiums and partially subsidizing maintenance costs. These incentives will enhance green roof construction on new and existing buildings with added social environmental benefits."

Moreover, according to the monitoring activities on the real estate market in Italy, ENEA (Italian National Agency for New Technologies, Energy and Sustainable Economic Development) 'Istituto per la Competitività (I-Com)' and the 'Federazione Italiana degli Agenti Immobiliari Professionisti (FIAIP)' show a strong correlation between the performing energy classes, potentially improved by the greening process of roofs, among other factors, and the trend in the real estate market. The general positive trend in 2019 showed a progressive reduction of buildings sold falling into the poorest energy class (G), which was around 40% for one-room and two-room apartments, 37% for three-room apartments, around 34% of single-family villas, and 24% for villas. Especially for these last two categories of homes, the improvement in the figure compared to the previous year is very positive. The signals coming from the renovated buildings' segment are also positive, with the percentage of buildings belonging to the best-performing energy classes (A +, A, and B) going from 10% in 2017 to 22% in 2018.

Social Benefits

Different social benefits have been associated with the green roof and rooftop utilization in general. Shafique et al. [31] explained that among the benefits of roof utilisation, green roofs, in particular, the relief from the concrete construction by introducing the green space in urban areas seems to be the most significant. This is due to the ability of green roofs to provide pleasant effects to the urban inhabitants by reducing the air and noise pollution. The authors also explained that the use of roof spaces attract the eyes and tried to connect people together, thanks to the possible plurality of activities. The use of roofs might also enhance the property values [31]. Oberndorfer et al. [86] also explained that: " … living roofs also provide aesthetic and psychological benefits for people in urban areas. Even when green roofs are only accessible as visual relief, the benefits may include relaxation and restoration, which can improve human health. Other uses for green roofs include urban agriculture: food production can provide economic and educational benefits to urban dwellers." Moreover, Williams et al. [87] stretched the concept of the use of roofs, as well as their greening process, as a factor of enjoyment, concentration improvement, crime reduction, and productivity and creativity improvement, as well as helping behaviors among citizens. Williams et al. [87] also reported that: " … Studies from Canada and Finland demonstrate that many visitors to case study green roofs report positive aesthetic and restorative experience. A post-occupancy evaluation of gardens in a hospital setting, including several roof top gardens, found psychological benefits for patients including emotion respite." The authors also explained that not only green roofs can provide such benefits, but also that the plurality of activities and functions that can be performed on roofs can enable a variety of socially positive effects related to the increase of well-being.

4.2. Smart Green Policies for Rooftop Renovation and Management

Berardi et al. [71] explained that a number of cities have already implemented successful policies to enable the diffusion of green roofs. The city of Tokyo has requested the implementation of green roofs in private buildings with built areas larger than 1000 m^2 and in public buildings with built areas larger than 250 m^2. Moreover, Berardi et al. [71] reported that, in 2014, Germany had in place a supporting program to facilitate the construction of 13.5 million m^2 of green roofs per year. Cities, such as Esslingen, offered 50% of the cost of green roofs back or Darmstadt Municipality allocated an economic benefit of maximum of euro 5000 for a green roof. Berardi et al. [71] also explained that: " ... In the cities of Bonn, Cologne and Mannheim, the allocated storm water fees are considerably reduced once new green roofs are built. Similar policies have been implemented in other countries such as Switzerland and Austria. In Basel, users are repaid 20% of the cost of a green roof. In Toronto, there have been specific policies to promote green roofs in buildings with the ratio of 50–70% of the entire building coverage. In Quebec, an economic incentive is provided per square meter implemented of green roofs."

The results of the study supported the definition of urban rules and regulation to improve the quality of life and livability, to promote a sustainable development of urban environment, and to identify effective energy policies for a more resilient city in the case of the Municipality of Turin. In particular, this work was formally included into the working papers of the Municipality of Turin. These documents are the technical knowledge-based support, on which the review process of the City Masterplan of the City of Turin is officially based [87].

5. Conclusions

This paper presented the work carried out within the Re-Coding project, a multidisciplinary exploration carried out by *FULL* and R3C at the Polytechnic of Turin, in collaboration with the Municipality of Turin. The study analyzed the role of regulation in fostering or hindering sustainable development and in supporting the Municipality of Turin to simplify, disambiguate, and redefine the local current environmental regulation system. The study particularly focused on the exploration of the role of the roof element, as fifth facades of buildings.

The study has evaluated the potential, feasibility, and impact of the rooftops' renovation for a district of Turin. The results of this investigation confirm that the use of green roofs and the production of energy with solar thermal collectors and photovoltaic panels mitigate the urban heat island effect, reducing energy consumption with environmental, economic, and social benefits.

In line with [4], the study supported the idea that simple regulation can support and foster the transformation of cities toward a more sustainable and resilient built environment. Further studies will need to be conducted with the aim of monitoring and updating the future transformations of our cities, based on regulation updates. To this end, the approach of this work will need to consider the call for an ongoing normative review system in line with technological advancement, as well as in line with [88] the ability of institutions to envision and promote a sustainable and resilient vision for our cities. As [8] explained, understanding the opportunities of sustainable codes and regulations can contribute to the development and management of our future sustainably built environment, as the built environment and its environmental governance can still be considered the reflection of the societal changes that we decide to address [89].

Author Contributions: Conceptualization: V.T., G.M., L.B., M.N., and M.R.; methodology: V.T. and G.M.; formal analysis: V.T. and G.M.; writing—original draft preparation: V.T., G.M., L.B., and M.N.; writing—review and editing, V.T., G.M., L.B., M.N., and M.R. All authors have read and agreed to the published version of the manuscript.

Funding: This research received no external funding.

References

1. Shen, B.; Ghatikar, G.; Lei, Z.; Li, J.; Wikler, G.; Martin, P. The role of regulatory reforms, market changes, and technology development to make demand response a viable resource in meeting energy challenges. *Appl. Energy* **2014**, *130*, 814–823. [CrossRef]
2. Lehnerer, A. *Grand Urban Rules*; 010 Publishers: Rotterdam, The Netherlands, 2009.
3. Marshall, S. *Urban Coding and Planning*; Routledge: London, UK; New York, NY, USA, 2011.

4. Moroni, S.; Buitelaar, E.; Sorel, N.; Cozzolino, S. Simple Planning Rules for Complex Urban Problems: Toward Legal Certainty for Spatial Flexibility. *J. Plan. Educ. Res.* **2018**, *40*, 320–331. [CrossRef]
5. Oswalt, P.; Overmeyer, K.; Misselwitz, P. *Urban Catalyst: The Power of Temporary Use*; DOM Publishers: Berlin, Germany, 2013.
6. Baum, M.; Christiaanse, K. *City as Loft: Adaptive Reuse as a Resource for Sustainable Urban Development*; Gta Verl: Zürich, Switzerland, 2012.
7. Slaughter, E.S. Implementation of construction innovations. *Build. Res. Inf.* **2000**, *28*, 2–17. [CrossRef]
8. Nigra, M.; Dimitrijevic, B. Is radical innovation in architecture crucial to sustainability? Lessons from three Scottish contemporary buildings. *Arch. Eng. Des. Manag.* **2018**, *14*, 272–291. [CrossRef]
9. Mutani, G.; Todeschi, V. An Urban Energy Atlas and Engineering Model for Resilient Cities. *Int. J. Heat Technol.* **2019**, *37*, 936–947. [CrossRef]
10. Mutani, G.; Todeschi, V.; Matsuo, K. Urban Heat Island Mitigation: A GIS-based Model for Hiroshima. *Instrum. Mes. Métrologie* **2019**, *18*, 323–335. [CrossRef]
11. Taha, H.; Sailor, D.; Municipal, S. *High-Albedo Materials for Reducing Building Cooling Energy Use. Energy*; U.S. Department of Energy Office of Scientific and Technical Information: Oak Ridge, TN, USA, 1992. [CrossRef]
12. Mutani, G.; Todeschi, V. The Effects of Green Roofs on Outdoor Thermal Comfort, Urban Heat Island Mitigation and Energy Savings. *Atmosphere* **2020**, *11*, 123. [CrossRef]
13. Mutani, G.; Todeschi, V.; Kampf, J.; Coors, V.; Fitzky, M. Building energy consumption modeling at urban scale: Three case studies in Europe for residential buildings. In Proceedings of the 2018 IEEE International Telecommunications Energy Conference (INTELEC), Turin, Italy, 7–11 October 2018; pp. 1–8.
14. Boghetti, R.; Fantozzi, F.; Kämpf, J.; Mutani, G.; Salvadori, G.; Todeschi, V. Building energy models with Morphological urban-scale parameters: A case study in Turin. In Proceedings of the BSA: Building Simulation Applications, Bozen-Bolzano, South Tyrol, Italy, 19–21 June 2019; ISBN 978-88-6046-176-6.
15. Mutani, G.; Todeschi, V. Space heating models at urban scale for buildings in the city of Turin (Italy). *Energy Procedia* **2017**, *122*, 841–846. [CrossRef]
16. Mutani, G.; Todeschi, V. Building energy modeling at neighborhood scale. *Energy Effic.* **2020**, *13*, 1353–1386. [CrossRef]
17. Mutani, G.; Todeschi, V.; Beltramino, S. Energy Consumption Models at Urban Scale to Measure Energy Resilience. *Sustainability* **2020**, *12*, 5678. [CrossRef]
18. Mutani, G.; Gabrielli, C.; Nuvoli, G. Energy Performance Certificates Analysis in Piedmont Region (IT). A New Oil Field Never Exploited Has Been Discovered. *Tec. Ital. J. Eng. Sci.* **2020**, *64*, 71–82. [CrossRef]
19. Zheng, Y.; Weng, Q.; Zheng, Y. A Hybrid Approach for Three-Dimensional Building Reconstruction in Indianapolis from LiDAR Data. *Remote. Sens.* **2017**, *9*, 310. [CrossRef]
20. Overwatch Textron Systems. *Feature Analyst 5.2 Reference Guide*; Overwatch Textron Systems: Austin, TX, USA, 2007.
21. Mutani, G.; Todeschi, V. Urban Building Energy Modeling: Hourly energy balance model of residential buildings at district scale. *J. Phys. Conf. Ser.* **2020**, *1599*. [CrossRef]
22. Mutani, G.; Todeschi, V. Low-Carbon Strategies for Resilient Cities: A Place-Based Evaluation of Solar Technologies and Green Roofs Potential in Urban Contexts. *Tec. Ital. J. Eng. Sci.* **2020**, *64*, 193–201. [CrossRef]
23. Botham-Myint, D.; Recktenwald, G.W.; Sailor, D.J. Thermal footprint effect of rooftop urban cooling strategies. *Urban Clim.* **2015**, *14*, 268–277. [CrossRef]
24. Santos, T.; Tenedório, J.A.; Gonçalves, J.A. Quantifying the City's Green Area Potential Gain Using Remote Sensing Data. *Sustainability* **2016**, *8*, 1247. [CrossRef]
25. Hong, T.; Lee, M.; Koo, C.; Jeong, K.; Kim, J. Development of a method for estimating the rooftop solar photovoltaic (PV) potential by analyzing the available rooftop area using Hillshade analysis. *Appl. Energy* **2017**, *194*, 320–332. [CrossRef]
26. Aparicio-Gonzalez, E.; Domingo-Irigoyen, S.; Snachez-Ostiz, A. Rooftop extension as a solution to reach nZEB in building renovation. Application through typology classification at a neighborhood level. *Sustain. Cities Soc.* **2020**, *57*, 102109. [CrossRef]
27. Urban, B.; Roth, K. *Guidelines for Selecting Cool Roofs*; United States Department of Energy: Washington, DC, USA, 2010; pp. 1–23.

28. Canto-Perello, J.; Martinez-Garcia, M.P.; Curiel-Esparza, J.; Martin-Utrillas, M. Implementing Sustainability Criteria for Selecting a Roof Assembly Typology in Medium Span Buildings. *Sustainability* **2015**, *7*, 6854–6871. [CrossRef]
29. Suter, I.; Maksimović, Č.; Van Reeuwijk, M. A neighbourhood—Scale estimate for the cooling potential of green roofs. *Urban Clim.* **2017**, *20*, 33–45. [CrossRef]
30. Shafique, M.; Kim, R.; Rafiq, M. Green roof benefits, opportunities and challenges—A review. *Renew. Sustain. Energy Rev.* **2018**, *90*, 757–773. [CrossRef]
31. Ziogou, I.; Michopoulos, A.; Voulgari, V.; Zachariadis, T. Implementation of green roof technology in residential buildings and neighborhoods of Cyprus. *Sustain. Cities Soc.* **2018**, *40*, 233–243. [CrossRef]
32. Tang, M.; Zheng, X. Experimental study of the thermal performance of an extensive green roof on sunny summer days. *Appl. Energy* **2019**, *242*, 1010–1021. [CrossRef]
33. Yang, J.; Bou-Zeid, E. Scale dependence of the benefits and efficiency of green and cool roofs. *Landsc. Urban Plan.* **2019**, *185*, 127–140. [CrossRef]
34. Dong, J.; Lin, M.; Zuo, J.; Tao, L.; Liu, J.; Sun, C.; Luo, J. Quantitative study on the cooling effect of green roofs in a high-density urban Area—A case study of Xiamen, China. *J. Clean. Prod.* **2020**, *255*, 120152. [CrossRef]
35. Dimitrijević, B. (Ed.) *Innovations for Sustainable Building Design and Refurbishment in Scotland*; Springer International Publishing: New York, NY, USA, 2013. [CrossRef]
36. Dimitrijevic, B.; Langford, D. Assessment Focus for More Sustainable Buildings. In Proceedings of the SUE-MoT International Conference on Whole Life Urban Sustainability and Its Assessment, Glasgow, UK, 27–29 June 2007.
37. He, Y.; Yu, H.; Ozaki, A.; Dong, N. Thermal and energy performance of green roof and cool roof: A comparison study in Shanghai area. *J. Clean. Prod.* **2020**, *267*, 122205. [CrossRef]
38. D'Orazio, M.; Di Perna, C.; Di Giuseppe, E. Green roof yearly performance: A case study in a highly insulated building under temperate climate. *Energy Build.* **2012**, *55*, 439–451. [CrossRef]
39. Del Barrio, E.P. Analysis of the green roofs cooling potential in buildings. *Energy Build.* **1998**, *27*, 179–193. [CrossRef]
40. Sanchez, L.; Reames, T.G. Cooling Detroit: A socio-spatial analysis of equity in green roofs as an urban heat island mitigation strategy. *Urban For. Urban Green.* **2019**, *44*, 126331. [CrossRef]
41. Hoelscher, M.-T.; Nehls, T.; Jänicke, B.; Wessolek, G. Quantifying cooling effects of facade greening: Shading, transpiration and insulation. *Energy Build.* **2016**, *114*, 283–290. [CrossRef]
42. Mahmoud, A.S.; Asif, M.; Hassanain, M.A.; Babsail, M.O.; Sanni-Anibire, M.O. Energy and Economic Evaluation of Green Roofs for Residential Buildings in Hot-Humid Climates. *Buildings* **2017**, *7*, 30. [CrossRef]
43. Ng, E.; Chen, L.; Wang, Y.; Yuan, C. A study on the cooling effects of greening in a high-density city: An experience from Hong Kong. *Build. Environ.* **2012**, *47*, 256–271. [CrossRef]
44. Cascone, S.; Catania, F.; Gagliano, A.; Sciuto, G. A comprehensive study on green roof performance for retrofitting existing buildings. *Build. Environ.* **2018**, *136*, 227–239. [CrossRef]
45. Zhang, L.; Deng, Z.; Liang, L.; Zhang, Y.; Meng, Q.; Wang, J.; Santamouris, M. Thermal behavior of a vertical green facade and its impact on the indoor and outdoor thermal environment. *Energy Build.* **2019**, *204*, 109502. [CrossRef]
46. Peng, L.L.; Jiang, Z.; Yang, X.; He, Y.; Xu, T.; Chen, S.S. Cooling effects of block-scale facade greening and their relationship with urban form. *Build. Environ.* **2020**, *169*, 106552. [CrossRef]
47. Krarti, M. Integrated Design of Energy Efficient Cities. In *Optimal Design and Retrofit of Energy Efficient Buildings, Communities, and Urban Centers*; Butterworth-Heinemann: Oxford, UK, 2018. [CrossRef]
48. Akbari, H.; Pomerantz, M.; Taha, H. Cool surfaces and shade trees to reduce energy use and improve air quality in urban areas. *Sol. Energy* **2001**, *70*, 295–310. [CrossRef]
49. Oleson, K.W.; Bonan, G.B.; Feddema, J. Effects of white roofs on urban temperature in a global climate model. *Geophys. Res. Lett.* **2010**, *37*. [CrossRef]
50. MacIntyre, H.; Heaviside, C. Potential benefits of cool roofs in reducing heat-related mortality during heatwaves in a European city. *Environ. Int.* **2019**, *127*, 430–441. [CrossRef]
51. Santamouris, M.; Ban-Weiss, G.; Osmond, P.; Paolini, R.; Synnefa, A.; Cartalis, C.; Muscio, A.; Zinzi, M.; Morakinyo, T.E.; Ng, E.; et al. Progress in Urban Greenery Mitigation Science—Assessment Methodologies Advanced Technologies and Impact On Cities. *J. Civ. Eng. Manag.* **2018**, *24*, 638–671. [CrossRef]

52. Akbari, H.; Levinson, R. Evolution of Cool-Roof Standards in the US. *Adv. Build. Energy Res.* **2008**, *2*, 1–32. [CrossRef]

53. Berdahl, P.; Akbari, H.; Rose, L. Aging of reflective roofs: Soot deposition. *Appl. Opt.* **2002**, *41*, 2355–2360. [CrossRef] [PubMed]

54. Akbari, H.; Berhe, A.; Levinson, R.; Graveline, S.; Foley, K.; Delgado, A. *Aging and Weathering of Cool Roofing Membranes*; Lawrence Berkeley National Lab: Berkeley, CA, USA, 2005.

55. Akbari, H.; Berdahl, P.; Levinson, R.M.; Wiel, S.; Miller, W.A.; Desjarlais, A. *Cool Color Roofing Materials*; Energy Technologies Area: Berkley, CA, US, 2006.

56. Kodysh, J.B.; Omitaomu, O.A.; Bhaduri, B.L.; Neish, B.S. Methodology for estimating solar potential on multiple building rooftops for photovoltaic systems. *Sustain. Cities Soc.* **2013**, *8*, 31–41. [CrossRef]

57. Schindler, B.Y.; Blaustein, L.; Lotan, R.; Shalom, H.; Kadas, G.J.; Seifan, M. Green roof and photovoltaic panel integration: Effects on plant and arthropod diversity and electricity production. *J. Environ. Manag.* **2018**, *225*, 288–299. [CrossRef] [PubMed]

58. Song, X.; Huang, Y.; Zhao, C.; Liu, Y.; Lu, Y.; Chang, Y.; Yang, J. An Approach for Estimating Solar Photovoltaic Potential Based on Rooftop Retrieval from Remote Sensing Images. *Energies* **2018**, *11*, 3172. [CrossRef]

59. Mutani, G.; Todeschi, V.; Novo, R.; Mattiazzo, G.; Tartaglia, A. *Le Isole Minori tra Sole, Mare e Vento*; ENEA: Rome, Italy, 2019. (In Italian)

60. Mutani, G.; Pastorelli, M.; De Bosio, F. A model for the evaluation of thermal and electric energy consumptions in residential buildings: The case study in Torino (Italy). In Proceedings of the International Conference on Renewable Energy Research and Applications (ICRERA), Glasgow, UK, 27–30 September 2015; pp. 1399–1404. [CrossRef]

61. Mutani, G.; Beltramino, S.; Forte, A. A Clean Energy Atlas for Energy Communities in Piedmont Region (Italy). *Int. J. Des. Nat. Ecodyn.* **2020**, *15*, 343–353. [CrossRef]

62. La Roche, P.; Berardi, U. Comfort and energy savings with active green roofs. *Energy Build.* **2014**, *82*, 492–504. [CrossRef]

63. Rakotondramiarana, H.T.; Ranaivoarisoa, T.F.; Morau, D. Dynamic Simulation of the Green Roofs Impact on Building Energy Performance, Case Study of Antananarivo, Madagascar. *Buildings* **2015**, *5*, 497–520. [CrossRef]

64. Costanzo, V.; Evola, G.; Marletta, L. Energy savings in buildings or UHI mitigation? Comparison between green roofs and cool roofs. *Energy Build.* **2016**, *114*, 247–255. [CrossRef]

65. Silva, C.M.; Gomes, M.G.; Silva, M. Green roofs energy performance in Mediterranean climate. *Energy Build.* **2016**, *116*, 318–325. [CrossRef]

66. Bevilacqua, P.; Bruno, R.; Arcuri, N. Green roofs in a Mediterranean climate: Energy performances based on in-situ experimental data. *Renew. Energy* **2020**, *152*, 1414–1430. [CrossRef]

67. Kumar, V.; Prasad, L. Performance Analysis of Three-sides Concave Dimple Shape Roughened Solar Air Heater. *J. Sustain. Dev. Energy Water Environ. Syst.* **2018**, *6*, 631–648. [CrossRef]

68. Gann, D.M.; Salter, A.J. Innovation in project-based, service-enhanced firms: The construction of complex products and systems. *Res. Policy* **2000**, *29*, 955–972. [CrossRef]

69. Bossink, B. *Eco-Innovation and Sustainability Management*; Informa UK Limited: London, UK, 2013.

70. Berardi, U.; GhaffarianHoseini, A.; GhaffarianHoseini, A. State-of-the-art analysis of the environmental benefits of green roofs. *Appl. Energy* **2014**, *115*, 411–428. [CrossRef]

71. Vijayaraghavan, K. Green roofs: A critical review on the role of components, benefits, limitations and trends. *Renew. Sustain. Energy Rev.* **2016**, *57*, 740–752. [CrossRef]

72. Castleton, H.; Stovin, V.; Beck, S.; Davison, J. Green roofs; building energy savings and the potential for retrofit. *Energy Build.* **2010**, *42*, 1582–1591. [CrossRef]

73. Boixo, S.; Diaz-Vicente, M.; Colmenar-Santos, A.; Castro, M.A. Potential energy savings from cool roofs in Spain and Andalusia. *Energy* **2012**, *38*, 425–438. [CrossRef]

74. Rosado, P.J.; Faulkner, D.; Sullivan, D.P.; Levinson, R. Measured temperature reductions and energy savings from a cool tile roof on a central California home. *Energy Build.* **2014**, *80*, 57–71. [CrossRef]

75. New, J.R.; Miller, W.A.; Huang, Y.J.; Levinson, R. Comparison of software models for energy savings from cool roofs. *Energy Build.* **2016**, *114*, 130–135. [CrossRef]

76. Seifhashemi, M.; Capra, B.R.; Milller, W.; Bell, J. The potential for cool roofs to improve the energy efficiency of single storey warehouse-type retail buildings in Australia: A simulation case study. *Energy Build.* **2018**, *158*, 1393–1403. [CrossRef]

77. Akbari, H.; Menon, S.; Rosenfeld, A. Global cooling: Increasing world-wide urban albedos to offset CO2. *Clim. Chang.* **2008**, *94*, 275–286. [CrossRef]

78. Sproul, J.; Wan, M.P.; Mandel, B.H.; Rosenfeld, A.H. Economic comparison of white, green, and black flat roofs in the United States. *Energy Build.* **2014**, *71*, 20–27. [CrossRef]

79. Chiri, G.M.; Achenza, M.; Canì, A.; Neves, L.; Tendas, L.; Ferrari, S. The Microclimate Design Process in Current African Development: The UEM Campus in Maputo, Mozambique. *Energies* **2020**, *13*, 2316. [CrossRef]

80. Teotónio, I.; Silva, C.M.; Cruz, C.O. Eco-solutions for urban environments regeneration: The economic value of green roofs. *J. Clean. Prod.* **2018**, *199*, 121–135. [CrossRef]

81. Bianchini, F.; Hewage, K. Probabilistic social cost-benefit analysis for green roofs: A lifecycle approach. *Build. Environ.* **2012**, *58*, 152–162. [CrossRef]

82. Corcelli, F.; Fiorentino, G.; Petit-Boix, A.; Rieradevall, J.; Gabarrell, X. Transforming rooftops into productive urban spaces in the Mediterranean. An LCA comparison of agri-urban production and photovoltaic energy generation. *Resour. Conserv. Recycl.* **2019**, *144*, 321–336. [CrossRef]

83. Mahdiyar, A.; Tabatabaee, S.; Abdullah, A.; Marto, A. Identifying and assessing the critical criteria affecting decision-making for green roof type selection. *Sustain. Cities Soc.* **2018**, *39*, 772–783. [CrossRef]

84. Wang, M.; Mao, X.; Gao, Y.; He, F. Potential of carbon emission reduction and financial feasibility of urban rooftop photovoltaic power generation in Beijing. *J. Clean. Prod.* **2018**, *203*, 1119–1131. [CrossRef]

85. Oberndorfer, E.; Lundholm, J.; Bass, B.; Coffman, R.R.; Doshi, H.; Dunnett, N.; Gaffin, S.; Köhler, M.; Liu, K.K.Y.; Rowe, B. Green Roofs as Urban Ecosystems: Ecological Structures, Functions, and Services. *Bioscience* **2007**, *57*, 823–833. [CrossRef]

86. Williams, K.J.; Lee, K.E.; Sargent, L.; Johnson, K.A.; Rayner, J.; Farrell, C.; Miller, R.E.; Williams, N.S. Appraising the psychological benefits of green roofs for city residents and workers. *Urban For. Urban Green.* **2019**, *44*. [CrossRef]

87. Barioglio, C.; Campobenedetto, D.; Marianna, N.; Barale, M.F.; Frassoldati, F.; Robiglio, M. *Re-Coding—Ripensare le Regole Della Città*; Department of Architecture (DAD), Politecnico di Torino: Torino, Italy, 2019; ISBN 978-88-85745-28-5. (In Italian)

88. Leach, M.; Rockström, J.; Raskin, P.; Scoones, I.; Stirling, A.; Smith, A.; Thompson, J.; Millstone, E.; Ely, A.; Arond, E.; et al. Transforming Innovation for Sustainability. *Ecol. Soc.* **2012**, *17*, 11–16. [CrossRef]

89. Mumford, L. *Stick and Stones, a Study on American Architecture and Civilization*; Dover Publications: New York, NY, USA, 1955.

Article

Building Energy an Simulation Model for Analyzing Energy Saving Options of Multi-Span Greenhouses

Adnan Rasheed [1,†], Cheul Soon Kwak [2,†], Hyeon Tae Kim [3] and Hyun Woo Lee [1,4,5,*]

[1]   Department of Agricultural Engineering, Kyungpook National University, Daegu 41566, Korea; adnanrasheed@knu.ac.kr
[2]   Smart Farm System Department, Culti Labs CO. LTD, Gangneung 25440, Korea; cskwak@cultilabs.com
[3]   Department of Bio-Industrial Machinery Engineering, Institute of Agricultural and Life Sciences, Gyeongsang National University, Jinju 52828, Korea; bioani@gnu.ac.kr
[4]   Institute of Agricultural Science & Technology, Kyungpook National University, Daegu 41566, Korea
[5]   Smart Agriculture Innovation Center, Kyungpook National University, Daegu 41566, Korea
*   Correspondence: whlee@knu.ac.kr; Tel.: +82-53-950-5736
†   These authors contributed equally to this work.

Received: 16 September 2020; Accepted: 29 September 2020; Published: 1 October 2020

Abstract: This study proposes a multi-span greenhouse Building Energy Simulation (BES) model using a Transient System Simulation (TRNSYS)-18 program. A detailed BES model was developed and validated to simulate the thermal environment in the greenhouse under different design parameters for the multi-span greenhouse. Validation of the model was carried out by comparing the results from computed and experimental greenhouse internal temperatures. The statistical analyses produced an R^2 value of 0.84, a root mean square error (RMSE) value of 1.8 °C, and a relative (r)RMSE value of 6.7%, showing good agreement between computed and experimental results. The validated proposed BES model was used to evaluate the effect of multi-span greenhouse design parameters including thermal screens, number of screens, orientation, covering materials, double glazing, north-wall insulation, roof geometry, and natural ventilation, on the annual energy demand of the greenhouse, subjected to Taean Gun (latitude 36.88° N, longitude 126.24° E), Chungcheongnam-do, South Korea winter and summer season weather conditions. Additionally, the proposed BES model is capable of evaluating multi-span greenhouse design parameters with daily and seasonal dynamic control of thermal and shading screens, natural ventilation, as well as heating and cooling set-points. The TRNSYS 18 program proved to be highly flexible for carrying out simulations under local weather conditions and user-defined design and control of the greenhouse. The statistical analysis of validated results should encourage the adoption of the proposed model when the underlying aim is to evaluate the design parameters of multi-span greenhouses considering local weather conditions and specific needs.

Keywords: greenhouse modeling; energy demand; thermal screen; shading screen

---

1. Introduction

Worldwide, greenhouses are used to provide favorable climate conditions to crops, where outdoor weather conditions are not suitable for crop production [1]. In severe climatic conditions, to achieve the desired temperature condition inside the greenhouse for optimal crop growth, different heating and cooling systems are used. Although good crop yields can be obtained in well-controlled micro-environments within the greenhouse, such systems tend to increase operational costs [2]. The heating cost in South Korean greenhouses is 30–40% of total production costs, which makes the greenhouse sector one of the most energy-intensive [3]. Ahmed et al. [4] reported that the heating costs in Canadian greenhouses are 15–20% of total production costs [1]. Globally, various technologies, especially renewable energy technologies, have been utilized to fulfill the energy demands of buildings, including those in the agricultural sector [5]. In addition to using different energy sources, energy saving techniques can be helpful in reducing energy costs [6]. Hence, the reduction in the energy demand of greenhouses is very important. Having an energy-efficient greenhouse can contribute to reducing its energy demand. Therefore, the design parameters of the greenhouse significantly affect its the energy consumption [4].

Several studies [7–14] performed by different researchers considered single-span greenhouse design analyses, including varying the greenhouse's shape and orientation and evaluating the saving potentials of each. In our previous studies, Rasheed et al. [15] and Ahmed et al. [1] evaluated the design parameters of a single-span greenhouse with fully closed conditions including the shape, orientation, span, and width of the single-span greenhouse. The studies found that the single-span gothic shaped, east–west orientated greenhouse covered with double glazing of Polymethylmethacrylate (PMMA) is the most energy efficient greenhouse design parameter. Two studies conducted by Lee at al., [16] and Ha et al. [17] analyzed the heating and cooling loads of the Venlo type and wide-span greenhouse roof geometry under different weather conditions in different parts of South Korea. The outcomes of the studies show that the wide-span greenhouse required less energy when compared with the venlo type multi-span greenhouse. Lee et al. [18] designed a Building Energy Simulation (BES) model of the multi-span greenhouse to predict the annual and maximum heating load at six different locations of the greenhouse in South Korea and evaluated the performance of the heating system. On behave of the maximum heating load at each location feasibility of the heating system was evaluated.

The most feasible and practical approach to selecting energy-efficient greenhouses according to local weather conditions is through simulations. There are many types of Building Energy Simulation (BES) software such as Energy Plus, RETScreen, and the Transient System Simulation Program (TRNSYS)-18. Baglivo et al. [19] suggested that the TRNSYS program is more convenient and easier to use for this kind of simulation including energy modeling of greenhouse control strategies. For this specific study, TRNSYS-18 software, which is widely used for the estimation of building energy load and energy system performance, was selected. Moreover, in agricultural greenhouses, the software demonstrates a very high level of flexibility in terms of improving the structure of varying case studies as well as carrying out their energy analysis [20].

The specific objectives of this study are as follows:

- To provide the necessary input data (Pre-processing) to carry out multi-span greenhouse simulation using the TRNSYS 18 program.
- To propose a Building Energy Simulation (BES) model using the TRNSYS-18 program to evaluate the multi-span greenhouse design parameters under dynamic weather conditions while considering the dynamic daily and seasonal control of screen and natural ventilation, heating and cooling set points.
- To study the effects of different design parameters on the performance of the multi-span greenhouse. These parameters are different angular orientations namely, 0°, 30°, 45°, 60°, 75°, 90°; covering material type namely, Polyethylene (PE), Polyvinylchloride (PVC), Horticulture glass (HG), Polymethylmethacrylate (PMMA), and Polycarbonate (PC); and double-glazing type (and thickness) namely, PC (4, 6, 8, 10, 16 mm) and PMMA (8, 16 mm). In addition, the effect of screen types (Ph-77, Ph-super, Luxous, Tempa, Polyester); natural ventilation, north-wall insulation, different roof geometries and wide-span greenhouses are considered.

The proposed BES model provides a tool for efficiently analyzing multi-span greenhouse design parameters while taking into account local weather conditions and crop needs. The proposed model allows the dynamic simulation of greenhouse systems and also enables the application of different control strategies.

2. Materials and Methods

2.1. Experimental Greenhouse for Validation

The 15-span experimental greenhouse located in Taean Gun (latitude 36.88° N, longitude 126.24° E, elevation 45 m asl), as shown in Figure 1, was used for validation purposes. The geographic location of the experimental greenhouse is presented in Figure 2. The multi-span experimental greenhouse had a rectangular based, Venlo type roof-shaped; the roof was covered with horticulture glass (HG-4 mm) and the side walls were covered with 16-mm polycarbonate (PC) material. Furthermore, inside of greenhouse, three thermal screens two Ph-super and one Ph-77 were under the roof, and one Ph-77 on the side of the roof was applied. The total floor area of the greenhouse was 7572.6 m^2, with length, width, and volume of the greenhouse as 120.2 m and 63 m, 57 m$^3 \cdot$m^{-2}, respectively, while each span width, ridge height, and eve height were 8 m, 7.48 m, and 6.5 m, respectively. The schematic diagram of the experimental greenhouse, as depicted in Figure 3a, shows the vertical view of one span while Figure 3b shows the horizontal view. The weather data, including air temperature, solar radiation, relative humidity, wind speed, and wind direction, were recorded at the site. The ambient air pressure data were not recorded at the experimental site but obtained from the Korean Meteorological Administration (KMA) from nearest weather station—Seosan (129) weather station, latitude 36.7° N, longitude 126.4° E. The weather data were recorded from January to December 2019, and the characteristics of the weather data are shown in Table 1. Furthermore, the weather data were used as an input in the Building Energy Simulation (BES) model to simulate the real weather condition of the experimental site. Moreover, the greenhouse inside temperature was monitored for validation purposes.

Figure 1. Experimental multi-span venlo type greenhouse at Taean Gun, Chungcheongnam-do, Korea.

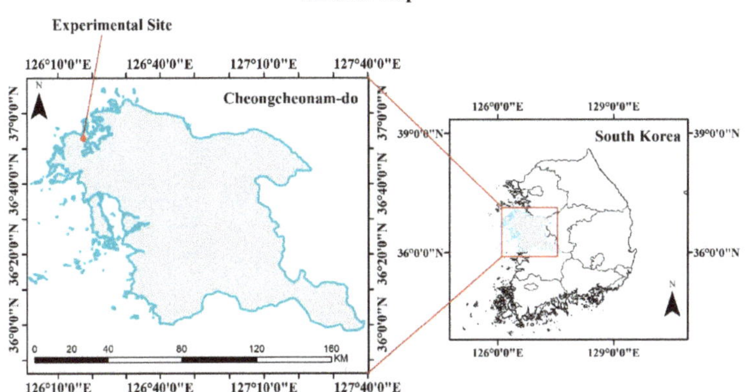

Figure 2. Geographic location of experimental greenhouse (latitude 36.88° N, longitude 126.24° E).

a

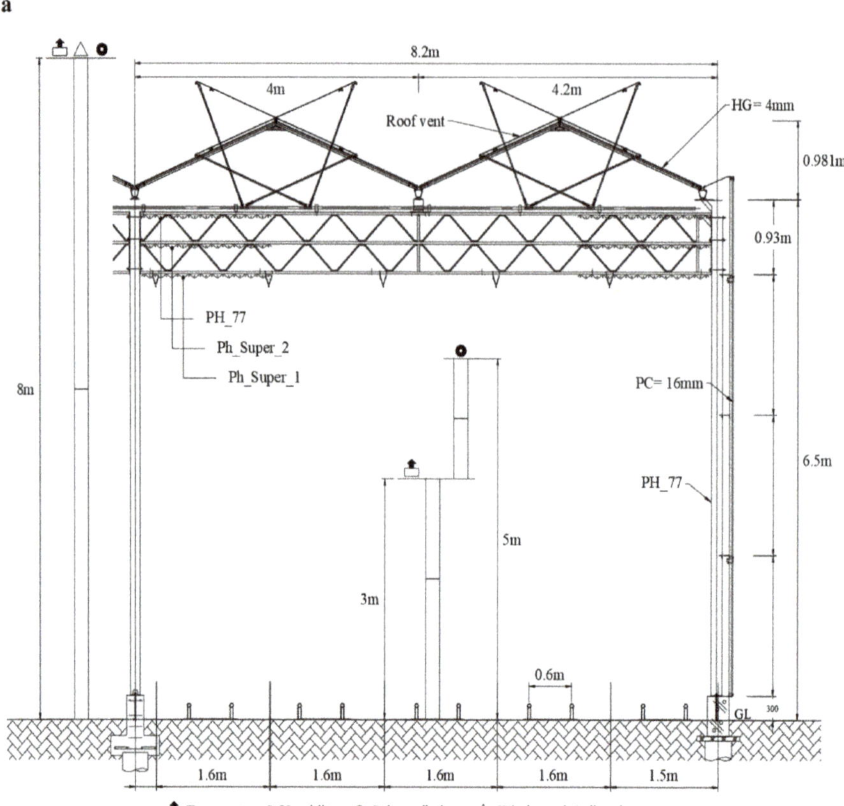

b

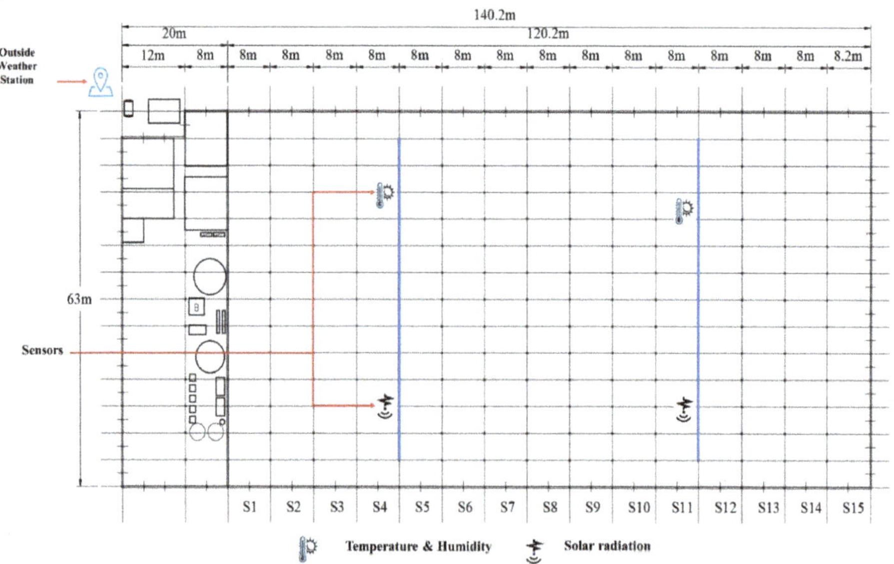

Figure 3. Experimental greenhouse dimensions (**a**) single span vertical view (**b**) horizontal view.

Table 1. Characteristics of the recorded weather data.

Weather Parameter	Unit	Time Interval	Sensor	Precision of Sensor	Data Recorded
Temperature	°C	1 min	IC, SHT75, SENSIRION	±0.3 °C	Field recorded
Relative humidity	%	1 min	IC, SHT75, SENSIRION	±1.8%	Field recorded
Solar radiation	$W{\cdot}m^{-2}$	1 min	ML-01C, Technox Inc	±2.0%	Field recorded
Wind speed	$m{\cdot}s^{-1}$	1 min	Model_Vantage Pro 2 6152CEU (Davis)	±5%	Field recorded
Wind direction	degree	1 min	Model_Vantage Pro 2 6152CEU (Davis)	±5%	Field recorded
Ambient pressure	hPa	1 min	PTB-220TS, VAISALA	±0.15%	KMA

2.2. BES Modeling

This study was conducted using TRNSYS 18, a BES modeling tool program as earlier stated. TRNSYS, which stands for "Transient System Simulation Program", is a versatile component-based program that provides tools for simulating both simple and complex energy flows in buildings. This program has been used for many different applications. Examples of simulations carried out using TRNSYS are those for energy systems research, energy building simulation, solar thermal processes, solar applications, ground-coupled heat transfer, geothermal heat pump systems, air-flow modeling, system calibration, hydrogen fuel cells, wind and Photovoltaic (PV) systems, and power plants [21].

The modeling process was completed in three parts, including, pre-processing, modeling, and simulation. The details of each are discussed below in the corresponding sub-sections. An overview of the complete modeling process is detailed in Figure 4.

2.2.1. Pre-Processing

The pre-processing of the BES model was carried out to prepare the input data using the following programs: Google SketchUp™ (3-D modeling program) software, and Transys3d (Add-on of TRNSYS program), Lawrence Berkeley National Laboratory (LBNL) Windows 7.7 software, Quick thermal conductivity meter (QTM-500). It is worth noting that the 3-D models for the multi-span greenhouse used in this study were prepared by using Google SketchUp™. Furthermore, Transys3d, an add-on of the TRNSYS program for Google SketchUp™, was used to prepare the IDF file of 3-D model. This was then imported as an input to TRNSYS-18. Figure 5 shows the 3-D model of the studied multi-span greenhouse with the Wide-span and the Venlo denoted as (a) and (b), respectively. The greenhouse coverings and screen materials are specific and cannot be used in conventional buildings; thermal screens are used inside the greenhouse during heating period to reduce heat loss to ambient weather to save heat energy and shading screens are used inside or outside of the greenhouse during the cooling period to minimize the solar heat gain inside the greenhouse which reduces greenhouse inside temperature and consequently reduces the cooling energy demand of the greenhouse, as well as different shading screens and strategies which effect the crop quality and yield [22]. To simulate the actual greenhouse indoor environment, greenhouse covers and thermal screens' physical, optical and thermal properties including thickness, thermal conductivity, solar, visible, and thermal, transmittance, emittance, and reflectance were used. For this specific purpose, the greenhouse covers' properties were taken from a study conducted by Valera et al. [23], with the summarized details presented in Table 2.

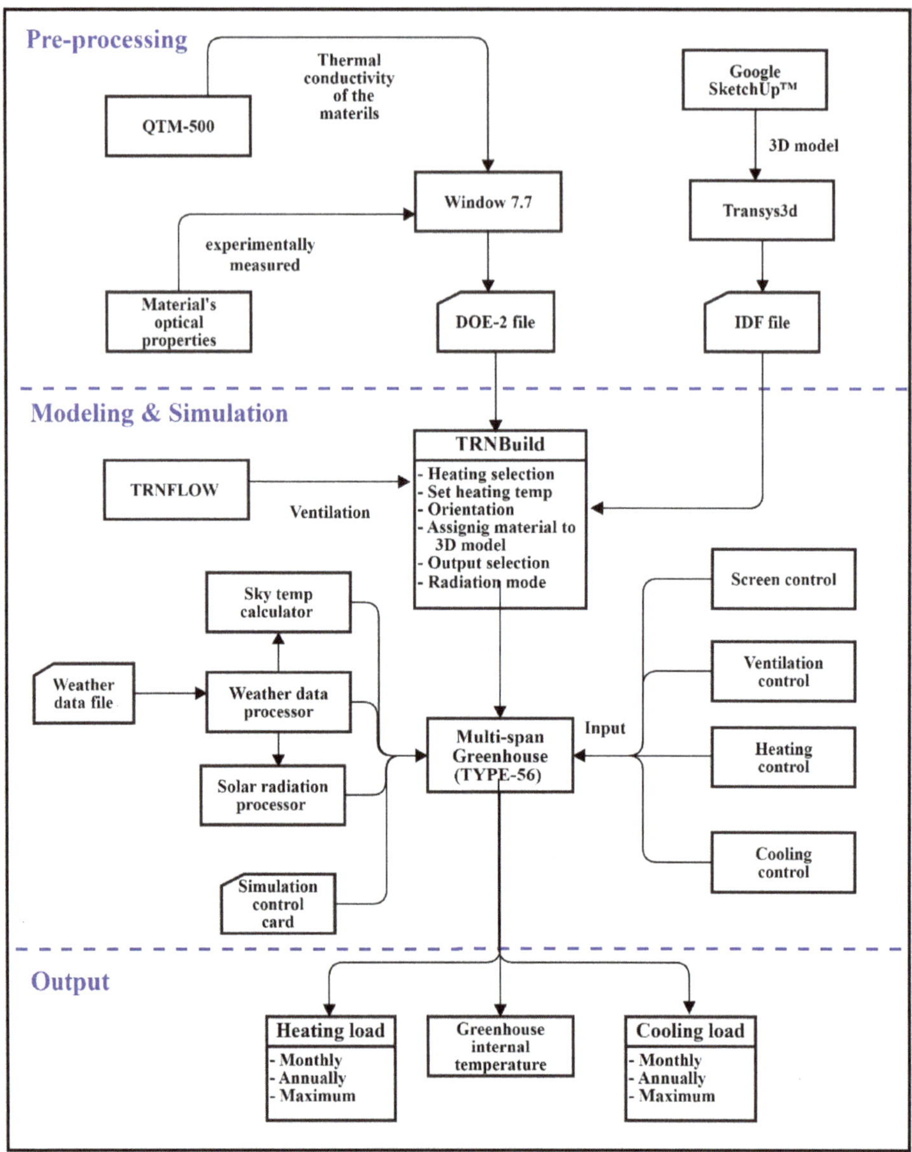

Figure 4. Flow diagram of the multi-span greenhouse Building Energy Simulation (BES) modeling.

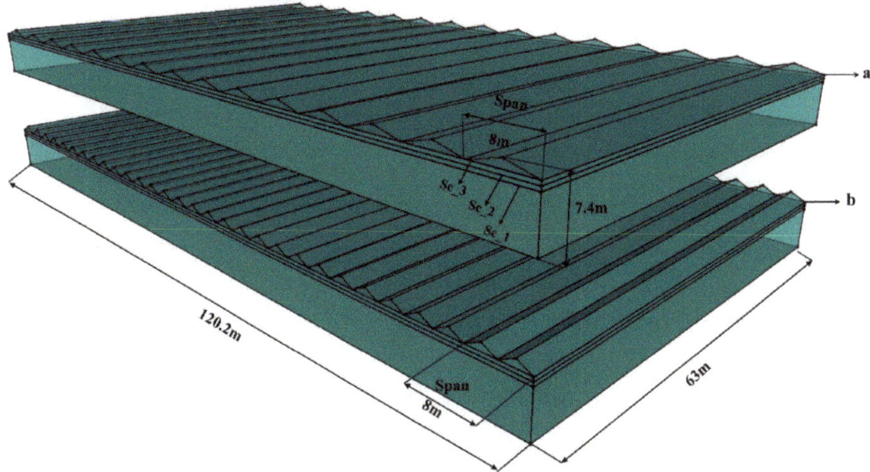

Figure 5. 3-D model of the studied multi-span greenhouse (**a**) Wide-span (**b**) Venlo.

Table 2. Physical, optical, and thermal properties of the greenhouse coverings.

Cover Characteristics	Greenhouse Materials				
	PC	HG	PE	PVC	PMMA
Solar radiation (300–2500 nm)					
Transmittance	0.78	0.89	0.86	0.91	0.78
Reflectance	0.14	0.08	0.10	0.07	0.14
Visible radiation (380–760 nm)					
Visible radiation transmittance	0.75	0.91	0.89	0.92	0.75
Visible radiation reflectance	0.15	0.08	0.08	0.14	0.15
Thermal radiation (2500–40,000 nm)					
Transmittance	0.02	0.10	0.18	0.06	0.02
Emmitance	0.89	0.90	0.79	0.62	0.89
Conductivity (W·m^{-1}·K^{-1})	0.19	0.76	0.33	0.13	0.19
Thickness (mm)	0.80	4.00	0.10	0.10	0.80

Furthermore, specific thermal and shading screens available in the South Korean market were used. It is worth noting that the properties of these materials are not in the literature, hence the need for characterization. The thermal conductivities of all the screens were measured by the QTM-500, a thermal conductivity measurement device; the details of the measuring process have earlier been published in Rasheed et al. [24]. Other properties of the screens, such as transmissivity, emissivity, and reflectivity for different types of radiation, were measured by the methodology described by Rafiq et al. [25]. A detailed summary of the physical, optical and thermal properties of the characterized screen are presented in Table 3. In addition, the greenhouse covers and screen material's properties were introduced into the LBNL Window 7.4 program to prepare a DOE-2 (readable by TRNBuild) file for each material. The DOE-2 is a specific file type which is used to introduce materials properties into the TRNSYS-18 for application on greenhouse 3-D models for further real-time simulation process.

Table 3. Physical, optical, and thermal properties of the screens.

Cover Characteristics	Greenhouse Materials				
	Ph-77	Ph-Super	Polyester	Luxous	Tempa
Solar radiation (300–2500 nm)					
Transmittance front	0.17	-	-	-	-
Transmittance back	0.17	-	-	-	-
Reflectance front	0.59	-	-	-	-
Reflectance back	0.51	-	-	-	-
Visible radiation (380–760 nm)					
Transmittance front	0.17	-	-	-	-
Transmittance back	0.17	-	-	-	-
Reflectance front	0.59	-	-	-	-
Reflectance back	0.51	-	-	-	-
Thermal radiation (2500–40,000 nm)					
Transmittance	0.20	0.38	0.02	0.38	0.01
Emmitance front	0.38	0.60	0.94	0.44	0.67
Emmitance back	0.48	0.60	0.94	0.44	0.67
Conductivity ($W \cdot m^{-1} \cdot K^{-1}$)	0.59	0.08	0.05	0.04	0.21
Thickness (mm)	0.40	0.30	0.40	0.22	0.25

2.2.2. Model Creation

The model was created using Simulation Studio, which is the main interface of the TRNSYS-18 program. TRNSYS-18 is a component-based program, and the simulation studio allows for the connection of all the components of the simulation together to create a model. The TRNSYS-18 Simulation studio for multi-span greenhouse model is presented in Figure 6 with the types used and their interconnections of this specific multi-span greenhouse BES model. Type-56 is first used to build the model component an interface is called TRNBuild. This is followed by the importation of IDF file for the 3-D model created by Transys3d of the multi-span greenhouse with DOE-2 files for all the cover and screen materials into TRNBuild. Application of the prepared materials into the 3-D TRNBuild is next. The TRNBuild deals with all the parameters and calculations of the greenhouse model, including solar radiation calculation on each surface of the greenhouse, convective, conductive, and radiative heat exchanges, heating and cooling set-points. The greenhouse ground properties, thermal conductivity, capacitance, and density were, 1.89 $kJ \cdot h^{-1} \cdot m^{-1} \cdot K^{-1}$, 1.5 $kJ \cdot kg^{-1} \cdot K^{-1}$, and 2000 $kg \cdot m^{-3}$, respectively, were also added in TRNBuild. The TRNFLOW, a tool for the calculation of the natural ventilation of the greenhouse, is also included in the TRNBuild interface. TRNBuild combines the thermal and ventilation model of the greenhouse. In simulation studio, the weather data is connected to Type-56 for the simulation of the greenhouse in the real situation. In the simulation studio different types were used to process the weather data and controllers for the dynamic day/night and seasonal control of natural vents opening and closing with inside temperature, screens control with outside solar radiations and heating and cooling set-points. A detailed description of all the specific components (Types) used in this modeling process is presented in Table 4.

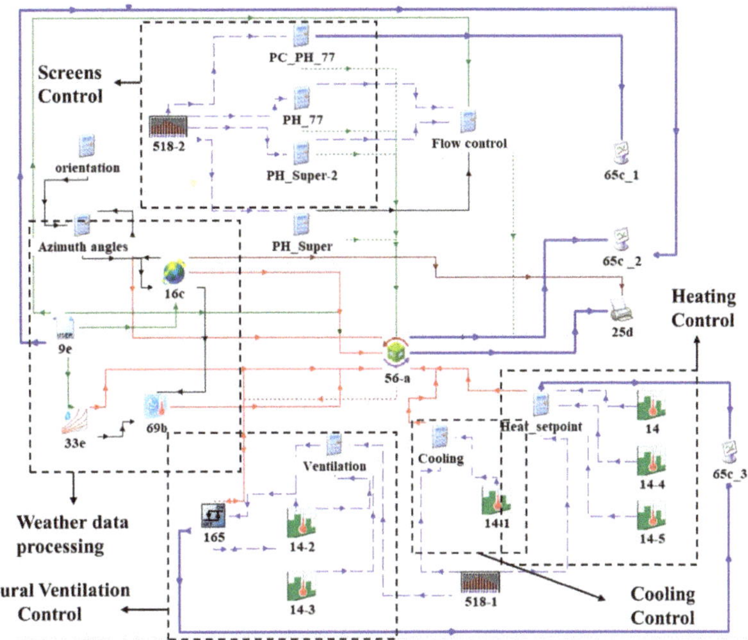

Figure 6. Transient System Simulation (TRNSYS) Simulation studio multi-span greenhouse model.

Table 4. Components of the greenhouse model in TRNSYS 18.

Component	Type	Description
Weather data reader	9e	This component serves the purpose of reading data at regular time intervals from a data file, converting it to a desired unit system.
Solar radiation processor	16c	Estimates total, beam, reflected, and diffused radiation on all greenhouse surfaces by utilizing total solar radiation on horizontal.
Sky temperature calculator	69b	Input: dew point temperature, beam and diffuse radiation on horizontal to calculate sky temperature for long-wave radiation exchange, the calculated sky temp.
Psychrometric chart	33e	This component calculates dew point temperature by utilizing inputs including, dry bulb temperature and humidity ratio.
Equation editor		This type was used to insert an equation. Equations were used to control, orientation, heat and cooling annual set-points, natural ventilation and screens deploy and retract.
Greenhouse building model with natural ventilation	56-a TRNFlow	1- This type is used to call TRNBuild to process the physical greenhouse 3D model. 2- TRNFlow is used to calculate natural ventilation air flow incorporation with a thermal model.
Controller	165	This type is a differential controller used to generate a control function which has a value of 0 and 1 for closed and open, respectively, used to control for deploying and retracting screens and opening and closing of vents for natural ventilation.
Monthly Function Scheduler	518	This type is used to generate, daily and monthly control values for the controller.
Time dependent forcing Function	14	It is used to control daily time varying set-points of heating and cooling
Printer	25d	This type was used to print results on the external user- provided file
Plotter	65c	This type was used to plot the results in an online plotter

2.2.3. Validation of the BES Model

To validate the proposed multi-span greenhouse BES model, the computed internal air temperature of the greenhouse was compared with those obtained experimentally using the same physical and operating conditions. Validation was carried out during a 10-day period in each of the summer and winter seasons of 2019 i.e., 20–29 August, and 1–10 December, respectively. These periods were chosen as operating conditions in greenhouses are different in these two periods. A summary of reference physical and operating conditions of the greenhouse during both time periods is given in Table 5. Furthermore, a statistical analysis of the validation results was performed in quantitative terms using the coefficient of determination (R^2), Equation (1); the root mean square error (RMSE), Equation (2); and the relative root mean square error (rRMSE), Equation (3). The R^2 value ranges between 0 and 1. A value nearer to 1 means the model is very accurate. The RMSE gives the standard deviation of the difference between computed and measured values. The rRMSR value is considered good if it is <10%, fair if it is <20%, and poor if it is >30%. They are mathematically defined as follows:

$$R^2 = 1 - \left[\frac{\sum_{i=0}^{n} \left(T_i^{exp} - T_i^{sim} \right)^2}{\sum_{i=0}^{n} \left(T_i^{exp} - T_i^{mean} \right)^2} \right] \tag{1}$$

$$RMSE = \sqrt{\frac{\sum_{i=0}^{n} \left(T_i^{exp} - T_i^{sim} \right)^2}{n}} \tag{2}$$

$$rRMSE = \frac{100}{T_i^{exp}} \left(\sqrt{\frac{\sum_{i=0}^{n} \left(T_i^{exp} - T_i^{sim} \right)^2}{n}} \right) \tag{3}$$

where T_i^{exp} is the experimentally obtained internal temperature of the greenhouse, T_i^{sim} is the simulated internal temperature of the greenhouse, T_i^{mean} is the mean of the experimental temperature, and n is the total number of observations.

2.2.4. Simulation

After validating our proposed model, further simulations were performed to investigate the effect of different greenhouse design parameters from an energy conservation point of view. Table 6 shows details of all the studied parameters. Firstly, an analysis was carried out to determine the effect of different orientations on the heating and cooling loads of the greenhouse that maintain the desired temperatures of 18 °C and 30 °C inside the greenhouse for heating and cooling, respectively. Further analysis was carried out with different covering materials and double glazing using the same greenhouse operating conditions that were used for validation (i.e., those presented in Table 5). Further analyses were performed to evaluate the effect of using no screen and using different thermal screens on the cooling and heating loads of the greenhouse. Comparisons were made to find the best solution. Moreover, the total annual energy load, including heating and cooling loads, was estimated with a fully closed and naturally ventilated greenhouse to predict the cooling energy reduction due to the use of natural ventilation. The effect of north wall insulation was estimated for the winter season only and heating load was calculated with and without insulation and a comparison was made. Finally, a venlo type and wide-span greenhouse heating and looing load was estimated.

Table 5. Summary of reference greenhouse conditions.

Parameter	Condition (a)	Condition (b)
Greenhouse type	Multi-span	Multi-span
No. of span	15	15
Roof type	Venlo	Venlo
Roof Glazing	HG-4 mm	HG-4 mm
Side Glazing	PC-16 mm	PC-16 mm
Dimension	120.2 m × 63 m × 7.5 m	120.2 m × 63 m × 7.5 m
Floor area	7572.6 m^2	7572.6 m^2
Orientation	North–South	North–South
Period	20–29 August 2019	1–10 December 2019
Screen	Shading (Ph-77)	Ph-77, Ph-Super
Screen control	Sunrise: 10:30 a.m.	Ph-77 retract (After sunrise, or Temp 10, or S.R 100 W) Ph-77 deploy (After sunset, or Temp 12, and S.R 100 W) Ph-Super_1 retract (After sunrise, or Temp 5, or S.R 50 W) Ph-Super_1 deploy (After sunset, or Temp 12, and S.R 50 W) Ph-Super_2 retract (After sunrise, or Temp 12, or S.R 150 W) Ph-Super_2 deploy (After sunset, or Temp 14, and S.R 150 W)
Natural ventilation	Roof vents	Roof vents
Natural vents control set point temperature	18 °C (00:00–04:00) 20 °C (04:00–10:00) 26 °C (10:00–18:00) 18 °C (18:00–24:00)	18 °C (00:00–05:30) 19 °C (05:30–16:15) 15 °C (16:15–21:30) 18 °C (21:30–24:00)
Heating	No	17 °C (00:00–16:15) 15 °C (16:15–21:30) 17 °C (21:30–24:00)

Table 6. Details of studied parameters.

Parameter		Test Condition					
Orientation		0 (E-W)	30	45	60	75	90 (N-S)
Covering material		PE	PVC	HG	PMMA		PC
Double glazing	PC (mm)	4	6	8	10		16
	PMMA (mm)		8			16	
No. of Screens	Thermal		1		2		3
	Shading				1		
Screen Materials	Ph-77		Ph-Super	Luxous	Tempa		Polyester
Natural ventilation			Fully closed		Naturally ventilated		
North-wall insulation			With		Without		
Roof shape			Venlo		Wide-span		

3. Results and Discussion

Figure 7a,b shows the summer and winter validation results. The computed results were obtained with the same physical and operating conditions as those of the experimental greenhouse, and are detailed in Table 5. Specifically, the internal greenhouse temperatures were compared. The statistical analysis of both validation results (given in Figure 7a,b) indicate R^2 values of 0.84 and 0.63; RMSE values of 1.8 °C and 1.3 °C; and rRMSE values of 6.7% and 7.4%, respectively. The R^2 value of 7a indicates that the model is accurate enough to calculate the inside temperature. In the case of the 7b, the value is a little less at 0.63, which is because the temperature fluctuations are much less. The RMSE value for the prediction of the greenhouse's internal temperature is less than those of the studies conducted by Ahamed et al. and Vanthoor et al. [26,27]. Their studies indicated that an rRMSE value less than 10% is reasonable. According to that, our model prediction is sufficiently accurate. The good agreement between the computed and experimental internal temperatures encourages the adoption of the proposed multi-span greenhouse BES model.

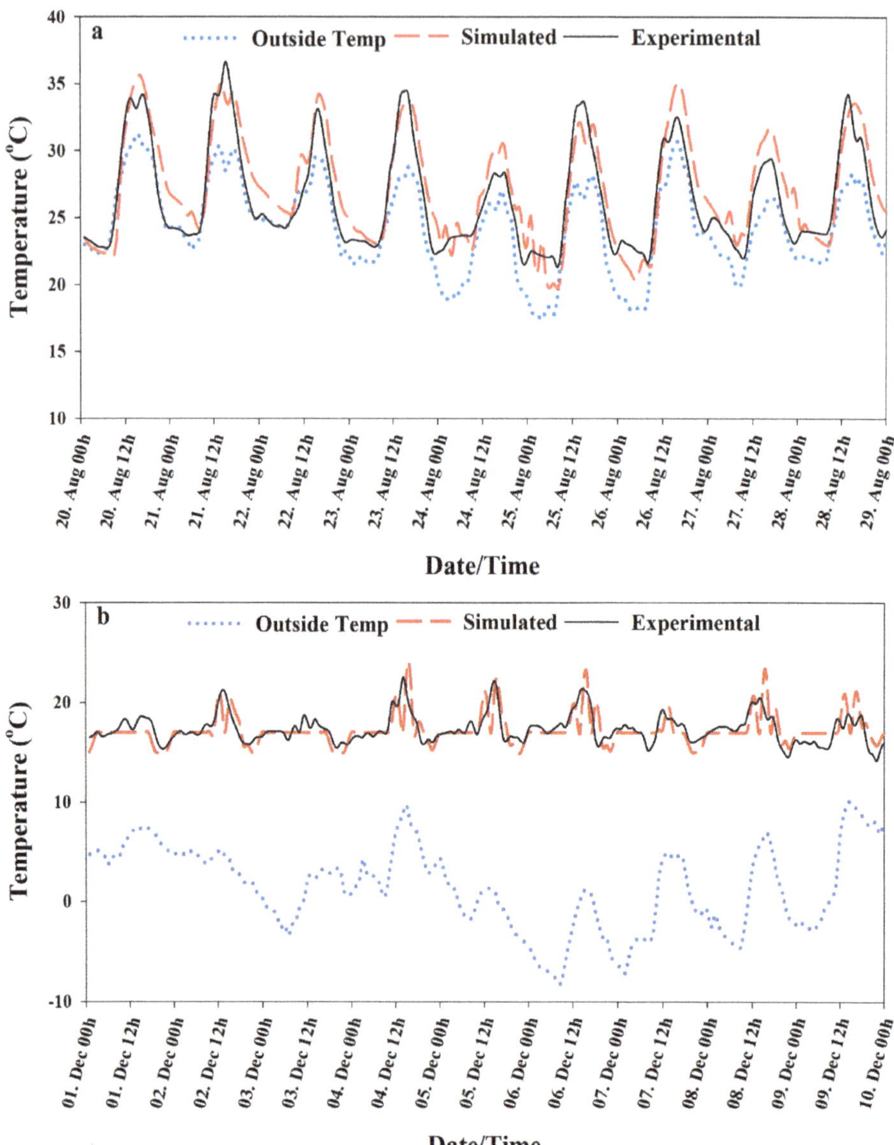

Figure 7. Computed versus measured internal air temperature of the greenhouse for validation in (**a**) summer, and (**b**) winter.

After successfully creating and validating the multi-span greenhouse BES model, simulations were carried out for all the multi-span greenhouse design parameters presented in Table 6 (simulation section). Further analyses were conducted with the same greenhouse description and operating conditions presented in Table 5. (validation section). Only heating and cooling set points were changed to 18 °C and 30 °C for winter and summer, respectively. Firstly, we evaluated the effect of using a number of screens in the greenhouse during heating and shading screens during cooling, and made a comparison. Figure 8 shows the monthly heating and cooling energy demand when thermal and shading screens were used. The months for heating and cooling were selected based on South Korean weather and crop needs. In the periods November to March we used heating, while cooling was used for June to September. The results showed significant heat energy savings during the winter period by using screens and compared with the no screen greenhouse case. Moreover, when using three screens inside the greenhouse, the heat energy demand was 70% and 40% lower than when the single and double layered screens were used. Normally, in greenhouses, one or two thermal screens are used. Geoola et al. [28,29], in two studies and our previous study (Rasheed et al. [30]) on the U-value of greenhouse cladding with thermal screens, reported that the use of thermal screens can reduce heating energy demand by 50–60%. Taki et al. [31] also reported a reduction in heating energy demand using thermal screens in comparison with cases without thermal screens. Their results are for the particular screens used in the current study, and can be different for other screen materials, as heat loss characteristics depend on the screen's properties. Furthermore, during summer months the shading screen showed 25% less cooling energy demand for the greenhouse. Therefore, during winter, thermal screens reduce heat loss to the ambient environment. Conversely, during summer, shading screens reduce the amount of solar radiation entering the greenhouse. This causes energy savings to be made and using screens in the greenhouse causes a volume decrease and hence a reduced energy demand. Ahmed et al. [32], who conducted a review on greenhouse shading for greenhouse energy savings improvement, and Abdel-Ghany et al. [33] who predicted the potential of different shading methods for greenhouses, also reported the same trend. In addition, Figure 9 shows the proposed model's capability in evaluating the thermal and shading screens effect on the greenhouse's energy demand under dynamic outside weather conditions and dynamic control of the screens (as in the real greenhouse).

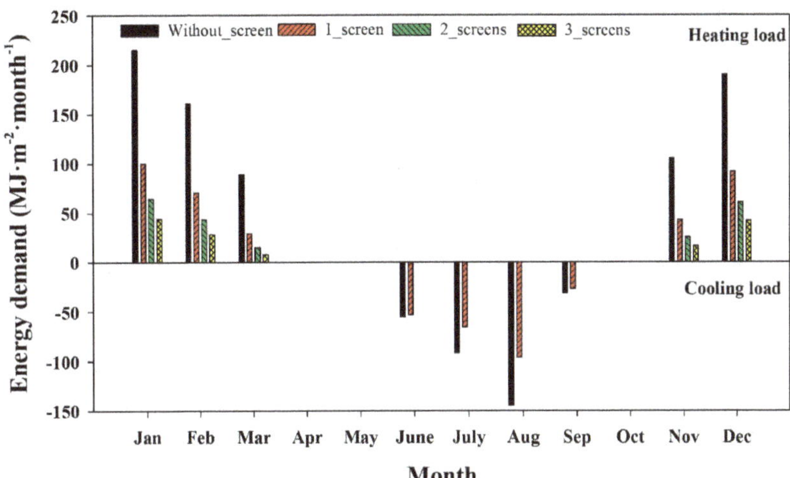

Figure 8. Comparison of heating and cooling loads of the greenhouse using thermal and shading screens.

Further analyses were conducted to estimate the maximum heating and cooling demands for the reference multi-span greenhouse with and without screens, and the results are presented in Table 7. The maximum heating demand occurred at 8:00 a.m. on 9 January 2019 when the outside temperature was −10.7 °C and the solar radiation was 0 kJ·h⁻¹·m⁻². The maximum cooling demand occurred at

3:00 PM on 7 July 2019 when the outside temperature was 35.7 °C and solar radiation was 3570 kJ·h^{-1}·m^{-2}. An estimation of the maximum heating and cooling demand of the greenhouse with fully controlled systems helps to design an energy providing facility for the greenhouse. Lee et al. [34,35] conducted two studies to estimate the maximum heating and cooling demand of a multi-span greenhouse to design a facility to fulfil its energy demand. They assessed the performance of the facility in fulfilling energy requirement in peak times.

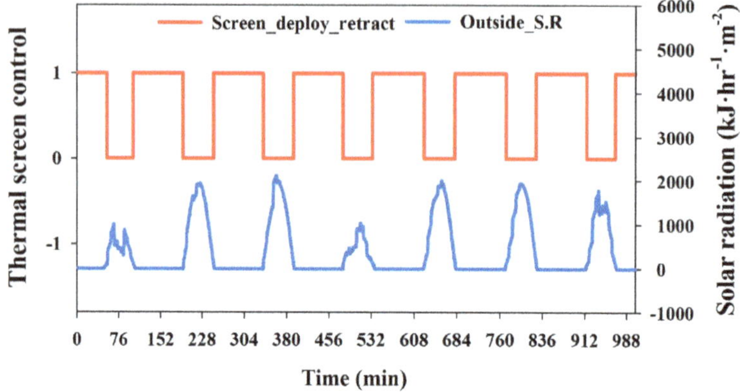

Figure 9. Screen control with outside solar radiation.

Table 7. Details of studied parameters.

	Maximum Heating Load		
	Load (MJ·h^{-1}·m^{-2})	Outside Temperature (°C)	Outside Solar Radiation (kJ·h^{-1}·m^{-2})
Without screen	0.65	−10.7	0
1 screen	0.46	−10.7	0
2 screens	0.41	−10.7	0
3 screens	0.34	−10.7	0
	Maximum Cooling Load		
Without shading screen	1.50	35.7	3570
with shading screen	1.18	35.7	3570

Different kinds of thermal screens are available in markets all over the world. Our screen analyses can help researchers and growers to choose the right screens according to their specific needs. For this reason, the heating energy demand of the different thermal screen materials available on the South Korean market were evaluated and compared. Their properties are described in Table 3. Figure 10 shows that Ph-77 combined with Ph-Super gives lowest least energy demand 139 MJ·m^{-2} compared with the others including Ph-77+Luxous, Ph-77+Polyester, Tempa+Ph-Super, Tempa+Ph-Luxous, Tempa+Polyester, which give demands of 142, 163, 145, 147, 171 MJ·m^{-2}, respectively. This is due to the fact that Ph-77 and Ph-super's emissivity values are lower than those of the others. Our previous study Rasheed et al. [28] and that of Ahamed et al. [30,36], on the sensitivity analyses of the effect of material properties on energy demand, confirmed these results.

Greenhouse covering material is also an important factor, as there are many covering materials available on the market. It has a direct effect on the solar gain and the energy requirement of the greenhouse, and choosing the most appropriate one according to crop needs can help in minimizing energy costs [15]. Figure 11 shows the monthly heating and cooling demand of the most commonly used greenhouse covers including, PE, PVC, HG, PC, PMMA. All the conditions were same as in the previous analysis and only side walls were replaced with different covering materials. A comparison was made of the estimated energy demand for all of them. The PC-16 mm material gave the smallest heating demand of, 20%, 19%, 7%, 4% during the heating months (Jan, Feb, Mar, Nov, Dec) than the other materials.. Furthermore, PE gave the smallest cooling demand of 2%, 7%, 5%, 4%, lower than the others during summer months (June, July, August, September).

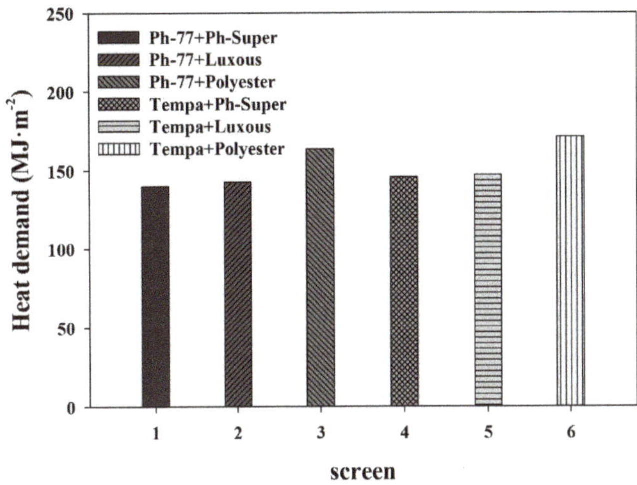

Figure 10. Heating demand of greenhouse using different thermal screens.

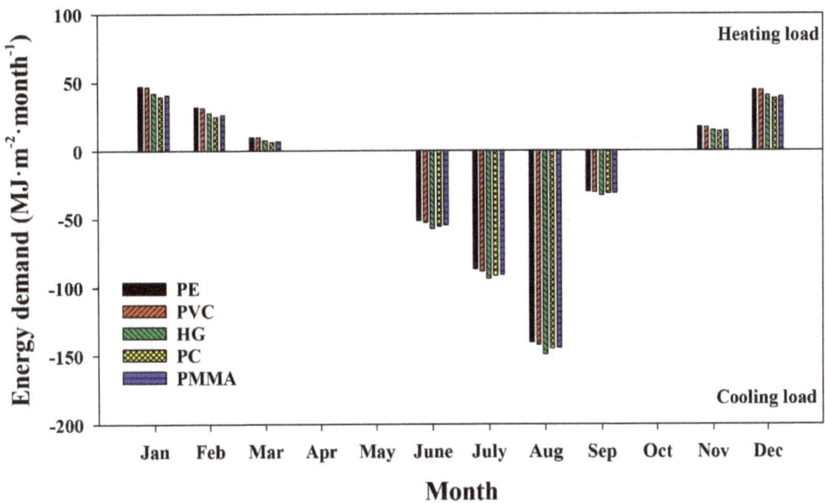

Figure 11. Monthly heating and cooling energy demand for the multi-span greenhouse using different greenhouse side wall covers. PE: Polyethylene, PVC: Polyvinyl chloride, HG: Horticulture Glass, PC: Poly carbonate, PMMA: Polymethylmethacrylate.

Further analyses were carried out for the investigation of the effect of side wall multi-glazing on heating and cooling energy demand for the multi-span greenhouse. The triple-, double-glazing materials PC, and PMMA with different thicknesses available on the South Korean market were evaluated. Figure 12 depicts the results for triple-layered PC-16, and double-layered PC (10, 8, 6, 4 mm), and PMMA (16, 10 mm) thickness. Each sheet of material was 0.8 mm thick, and hence the total thickness depends on the air gap between two sheets. The results in Figure 12a showed that the smallest heating demand is for the PC-16 mm material, and that the heating demand increased significantly for lower thicknesses. This is due to the fact that a thicker material serves better in preventing heat loss. Moreover, PC-16 mm gave 4% less heating demand than PMMA-16 mm. In Figure 12b, the cooling demand trend is the inverse of the heating demand and the lower thickness of PC-4 mm showed the smallest cooling demand. This is due to the fact that a higher thickness will resist heat loss to the ambient weather, thereby causing the internal temperature of the greenhouse to increase more than with the thinner material. Consequently, the cooling demand is increased. Figure 12c shows the total annual energy demand including heating and cooling for all the selected double glazing. The PC-16 mm material is more energy efficient than the double-glazed material.

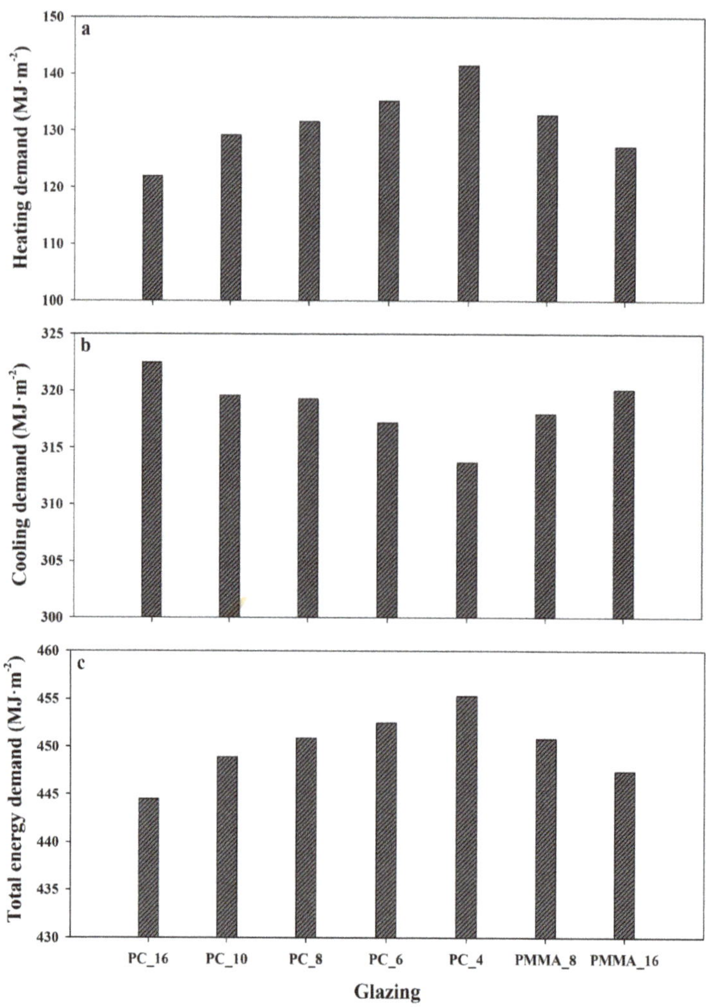

Figure 12. Annual energy demand of greenhouse with double-glazing (**a**) Heating (**b**) Cooling, (**c**) Total.

Greenhouse orientation has a significant effect on passive energy saving techniques. The effect of different greenhouse orientations was studied with the physical operating conditions described in Table 5. The total annual energy demand, including heating and cooling, was estimated for each orientation and the results are shown in Figure 13. The results indicate that the E–W orientation has the smallest annual energy requirement. This is due to the fact that, in winter, the E–W orientation receives more solar radiation than the other orientations, and the reverse happens in summer. Figure 14 shows the average daily solar energy gain inside the greenhouse during January (winter) and June (summer). The result indicates that, in January, the average daily solar gain of the E–W (0°) orientation was 11%, 14%, 15%, 12%, and 7% higher than the 15°, 30°, 45°, 60°, 75°, and 90° (N–S) orientations, respectively. Moreover, during June, the E–W (0°) orientation received 7%, 11%, 13%, 15%, and 15% less average solar gain than that of the 15°, 30°, 45°, 60°, 75°, and 90° (N–S) orientations, respectively. A similar study conducted for energy efficient design of a multi-span greenhouse also confirmed this trend [1].

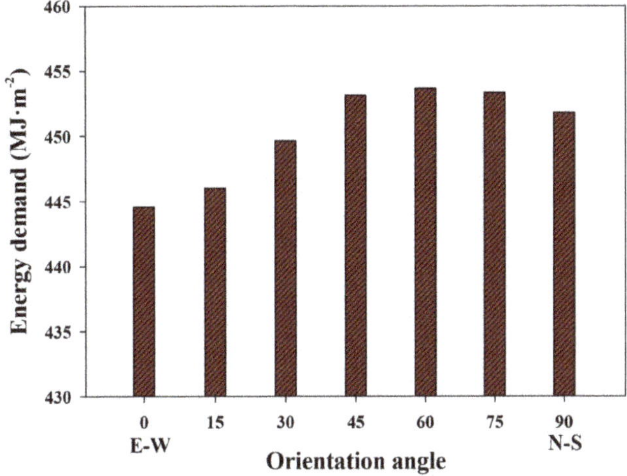

Figure 13. Annual energy demand for the greenhouse with different orientations.

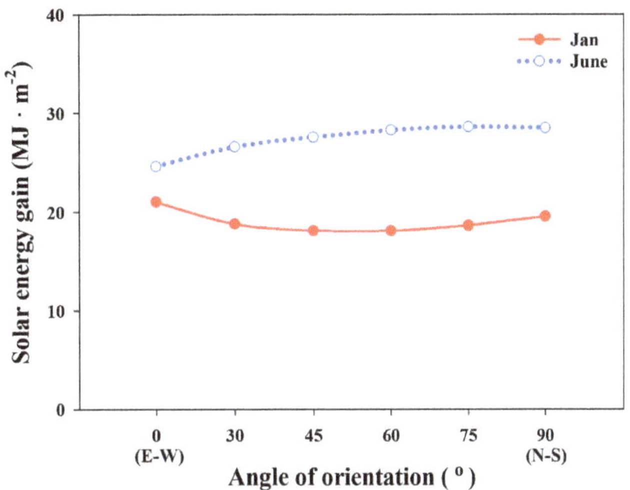

Figure 14. Annual energy demand for the greenhouse with different orientations.

Natural ventilation is widely used to decrease the internal temperature of greenhouses, especially in the summer. In this study, we evaluated the effect of natural ventilation on the cooling demand of the multi-span greenhouse. We estimated a monthly cooling demand of a fully closed and naturally ventilated greenhouse and the results are shown in Figure 15. Again, all physical and operating conditions are same ones presented in Table 5. The results indicate a 50% reduction in cooling demand during the whole summer season using natural ventilation combined with cooling. Figure 16 shows the internal temperature of the greenhouse with natural ventilation and a fully-closed condition. It can be seen that in the fully-closed condition with no ventilation, the greenhouse's internal temperature increases to 60 °C. However, with natural ventilation, it reduces to 40 °C, which causes a reduction in the cooling demand in the natural ventilation condition. The TRNSYS model successfully described the natural ventilation effect combined with a greenhouse thermal model.

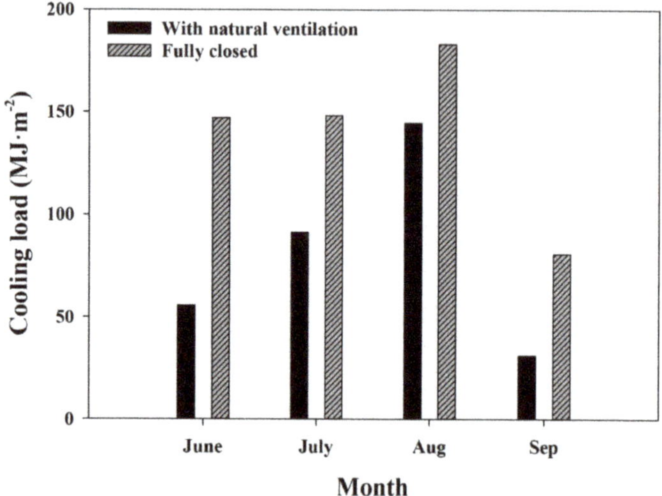

Figure 15. Comparison of monthly cooling demand of the greenhouse with natural ventilation and fully closed conditions.

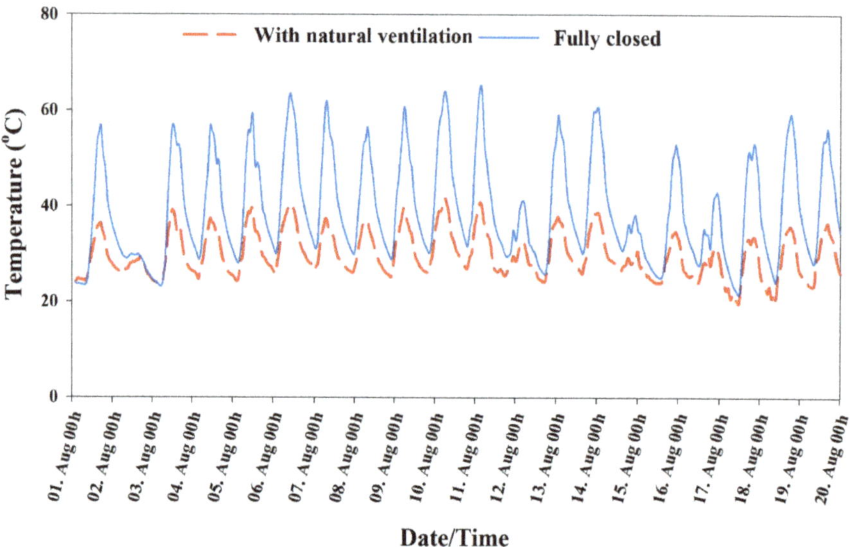

Figure 16. Comparison of greenhouse internal temperature with natural ventilation and fully closed conditions.

In the northern hemisphere, the sun stays at the south side and north wall of the greenhouse and contributes less to the solar heat gain inside the greenhouse. Other studies reported that the use of an opaque north wall can reduce 25–30% of the heating requirement in winter when compared to using a transparent north wall during the daytime. Therefore, it is recommended that insulation should be used on the north wall of the greenhouse to reduce heat loss, especially in colder regions during winter [4,37]. Commonly, multi-layered thermal screens are used for north wall insulation in South Korea. To simulate the specific insulation, we need to have the physical, optical and thermal properties of that material as described in Table 3. Due to lack of availability of these properties, for this specific reason (to show the model's capability) the available Ph-77 material was used as north wall insulation. Figure 17 shows the comparison between the heating demand of the multi-span greenhouse with north wall insulation and without insulation (transparent wall). The conditions are the same as those in aforementioned analyses. The overall results indicate that a 5% reduction in heating demand during the whole winter season was achieved using north wall insulation. The reduction of the heating demand purely depends on the material used for the north wall. Moreover, the specific energy saving is different as location and available solar radiation are different and the insulation material can also affect energy savings. Other similar studies for greenhouse design parameters, conducted by Gupta and Chandra [7], and Ahamed et al. [4], also confirmed this trend.

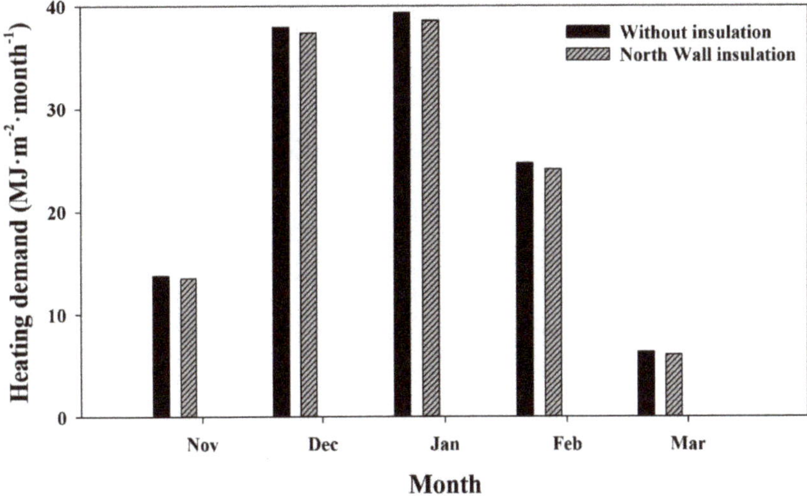

Figure 17. Comparison of monthly heating demand of the greenhouse with and without north wall insulation.

Greenhouse roof geometry has a huge influence on the annual energy demand of the greenhouse, as solar radiation received by the greenhouse influences its internal temperature [12]. Two multi-span greenhouse roof geometries (venlo and wide-span) were selected in order to compare their annual energy demands. The designs were selected according to South Korean standards for multi-span greenhouses. The results of both greenhouses' annual energy demand are shown in the Figure 18. The results indicate that during the winter season the venlo type greenhouse requires 25% less heating energy than the wide-span. In contrast, during the cooling season, the wide-span greenhouse requires 35% less cooling energy than the venlo type. This is due to the fact that the solar energy gain of the greenhouse with the wide-span roof geometry is less than that of the venlo type greenhouse. Figure 19 shows the monthly solar energy gain of both greenhouses. The venlo type greenhouse received more solar energy than that of the wide-span greenhouse. The solar gain inside the greenhouse was different due to the fact that the wide-span greenhouse has a lower angle of incidence—which causes low transmission—than that of the venlo type greenhouse with high angle of incidence—which causes high transmission. A study was conducted by Ha et al. [17] with the same type of greenhouse but the outcomes of the study show the wide-span greenhouse needs less annual energy than that of venlo type greenhouse. In that study, the greenhouse dimensions were not the same, so comparisons cannot be made. In our study, we used the same greenhouse dimension with but with a different roof geometry as shown in Figure 5.

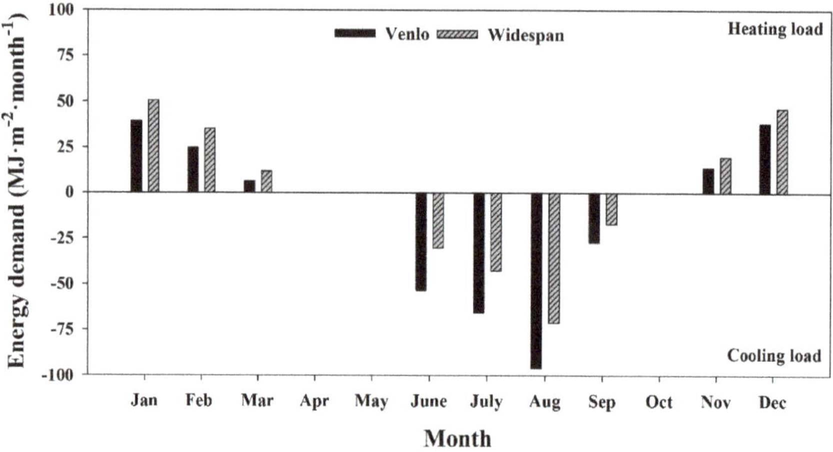

Figure 18. Comparison of monthly heating and cooling demand for selected greenhouse roof geometries.

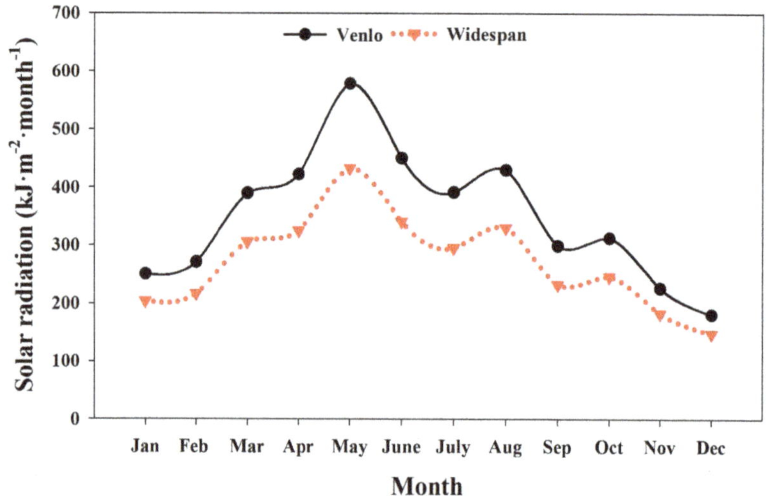

Figure 19. Comparison of monthly solar radiation gain for selected greenhouse roof geometries.

4. Conclusions

This study proposed a multi-span greenhouse BES model utilizing the TRNSYS 18 program. We detailed the creation and validation of the BES model for multi-span greenhouses to simulate the thermal environment of the greenhouse under different design parameters. The model was used to evaluate the effect of the parameters on the annual energy demand of the greenhouse. The proposed BES model is capable of evaluating the multi-span greenhouse's design parameters with daily and seasonal dynamic control of thermal and shading screens, natural ventilation, as well as heating and cooling set-points. The TRNSYS 18 program shows a high level of flexibility to carry out simulations under local weather conditions and can handle user-defined design and control of the greenhouse. The statistical analysis of the validated results encourages the adoption of the proposed model when the underlaying aim is to evaluate the multi-span greenhouse design parameters under local weather conditions and specific needs. The main findings of the study are as follows:

1- In winter, it is best use to three layers of thermal screens, as the analysis showed a 70% and 40% heat energy reduction compared with single and double layers of thermal screens. Moreover, the use of a shading screen in summer can reduce cooling energy demand by up to 25%.

2- The maximum heating loads of a multi-span greenhouse with conditions of without thermal screen, one screen, two screens, and three screens, were 0.65, 0.46, 0.41, and 0.344 $MJ \cdot h^{-1} \cdot m^{-2}$, respectively. Moreover, the maximum cooling loads without and with shading screen were, 1.5, and 1.18 $MJ \cdot h^{-1} \cdot m^{-2}$, respectively.

3- The Ph-77 thermal screen combined with Ph-super is more energy efficient and reduces heating energy demand during winter.

4- The PC-16mm showed the smallest heating demands of 20%, 19%, 7%, and 4% during the heating months than the other materials and thicknesses studied. Furthermore, PE showed the smallest cooling demands of 2%, 7%, 5%, and 4%. This was lower than the others studied.

5- PC-16 mm is a more energy efficient double-glazed material than the other materials studied.

6- The E–W orientation has smallest annual energy requirement when compared with the other orientations considered. The result indicates that in January, the average daily solar gain of the E–W (0°) orientation was, 11%, 14%, 15%, 12%, and 7%; and it was higher than the 15°, 30°, 45°, 60°, 75°, and 90° (N–S) orientations, respectively. Moreover, during June, the E–W (0°) orientation received 7%, 11%, 13%, 15%, and 15% less average solar gain than that of the 15°, 30°, 45°, 60°, 75°, and 90° (N–S) orientations, respectively.

7- A 50% reduction in cooling demand was experienced during the whole summer season using natural ventilation combined with cooling.

8- Using north wall insulation during the day, a 5% reduction in heating energy demand was achieved during whole winter season.

9- The Venlo type greenhouse requires 25% less heat energy compared with the wide-span type and vice versa in the cooling season. The wide-span greenhouse requires 35% less cooling energy than the venlo type. Moreover, the solar energy gain of a wide-span greenhouse roof geometry is less than that of the venlo type greenhouse.

Author Contributions: Conceptualization, A.R., H.W.L., C.S.K. and H.T.K.; Methodology, A.R., C.S.K., and H.W.L.; Software, A.R., C.S.K.; Validation, C.S.K. and H.T.K.; Investigation, A.R., C.S.K.; Resources, W.H.L.; Writing—Original Draft Preparation, A.R. and C.S.K.; Writing—Review and Editing, C.S.K., H.T.K. and H.W.L.; Supervision, H.W.L. All authors have read and agreed to the published version of the manuscript.

Funding: This work was supported by Korea Institute of Planning and Evaluation for Technology in Food, Agriculture, Forestry and Fisheries (IPET) through Agriculture, Food and Rural Affairs Convergence Technologies Program for Educating Creative Global Leader, funded by Ministry of Agriculture, Food and Rural Affairs (MAFRA) (717001-7). This research was supported by Basic Science Research Program through the National Research Foundation of Korea (NRF) funded by the Ministry of Education (NRF-2019R1I1A3A01051739).

References

1. Ahamed, M.S.; Guo, H.; Tanino, K. Energy-efficient design of greenhouse for canadian prairies using a heating simulation model. *Int. J. Energy Res.* **2018**, *42*, 2263–2272. [CrossRef]

2. Vadiee, A.; Martin, V. Solar blind system- solar energy utilization and climate mitigation in glassed buildings. *Energy Procedia* **2014**, *57*, 2023–2032. [CrossRef]

3. Yang, S.H.; Lee, C.G.; Lee, W.K.; Ashtiani, A.A.; Kim, J.Y.; Lee, S.D.; Rhee, J.Y. Heating and cooling system for utilization of surplus air thermal energy in greenhouse and its control logic. *J. Biosyst. Eng.* **2012**, *37*, 19–27. [CrossRef]

4. Ahamed, M.S.; Guo, H.; Tanino, K.J.B.E. Energy saving techniques for reducing the heating cost of conventional greenhouses. *Biosyst. Eng.* **2019**, *178*, 9–33. [CrossRef]

5. Xaman, J.; Hernandez-Perez, I.; Arce, J.; Alvarez, G.; Ramirez-Davila, L.; Noh-Pat, F. Numerical study of earth-to-air heat exchanger: The effect of thermal insulation. *Energy Build.* **2014**, *85*, 356–361. [CrossRef]

6. Rasheed, A.; Lee, J.W.; Lee, H.W. Development of a model to calculate the overall heat transfer coefficient of greenhouse covers. *Span. J. Agric. Res.* **2017**, *15*, 1–11. [CrossRef]

7. Gupta, M.J.; Chandra, P. Effect of greenhouse design parameters on conservation of energy for greenhouse environmental control. *Energy* **2002**, *27*, 777–794. [CrossRef]

8. Gupta, A.; Tiwari, G. Performance evaluation of greenhouse for different climatic zones of india. *J. Sol. Energy Soc. India* **2002**, *12*, 45.

9. Kumari, N.; Tiwari, G.; Sodha, M. Performance evaluation of greenhouse having passive or active heating in different climatic zones of india. *Agric. Eng. Int. CIGR J.* **2007**, *9*, 1–19.

10. Sethi, V.P. On the selection of shape and orientation of a greenhouse: Thermal modeling and experimental validation. *Sol. Energy* **2009**, *83*, 21–38. [CrossRef]

11. Çakır, U.; Şahin, E. Using solar greenhouses in cold climates and evaluating optimum type according to sizing, position and location: A case study. *Comput. Electron. Agric.* **2015**, *117*, 245–257. [CrossRef]

12. Djevic, M.; Dimitrijevic, A. Energy consumption for different greenhouse constructions. *Energy* **2009**, *34*, 1325–1331. [CrossRef]

13. Mobtaker, H.G.; Ajabshirchi, Y.; Ranjbar, S.F.; Matloobi, M. Solar energy conservation in greenhouse: Thermal analysis and experimental validation. *Renew. Energy* **2016**, *96*, 509–519. [CrossRef]

14. El-Maghlany, W.M.; Teamah, M.A.; Tanaka, H. Optimum design and orientation of the greenhouses for maximum capture of solar energy in north tropical region. *Energy Convers. Manag.* **2015**, *105*, 1096–1104. [CrossRef]

15. Rasheed, A.; Lee, J.; Lee, H. Development and optimization of a building energy simulation model to study the effect of greenhouse design parameters. *Energies* **2018**, *11*, 2001. [CrossRef]

16. Lee, S.-B.; Lee, I.-B.; Homg, S.-W.; Seo, I.-H.; Bitog, P.J.; Kwon, K.-S.; Ha, T.-H.; Han, C.-P. Prediction of greenhouse energy loads using building energy simulation (bes). *J. Korean Soc. Agric. Eng.* **2012**, *54*, 113–124.

17. Ha, T.; Lee, I.-B.; Kwon, K.-S.; Hong, S.-W. Computation and field experiment validation of greenhouse energy load using building energy simulation model. *Int. J. Agric. Biol. Eng.* **2015**, *8*, 116–127.

18. Lee, S.-N.; Park, S.-J.; Lee, I.-B.; Ha, T.-H.; Kwon, K.-S.; Kim, R.-W.; Yeo, U.-H.; Lee, S.-Y. Design of energy model of greenhouse including plant and estimation of heating and cooling loads for a multi-span plastic-film greenhouse by building energy simulation. *Prot. Hortic. Plant Fact.* **2016**, *25*, 123–132. [CrossRef]

19. Baglivo, C.; Mazzeo, D.; Panico, S.; Bonuso, S.; Matera, N.; Congedo, P.M.; Oliveti, G. Complete greenhouse dynamic simulation tool to assess the crop thermal well-being and energy needs. *Appl. Therm. Eng.* **2020**, *179*, 115698. [CrossRef]

20. Carlini, M.; Castellucci, S. Modelling and simulation for energy production parametric dependence in greenhouses. *Math. Probl. Eng.* **2010**, *2010*. [CrossRef]

21. Klein, S.A. *Trnsys, A Transient System Simulation Program*; Solar Energy Laborataory, University of Wisconsin: Madison, WI, USA, 2012.

22. Castronuovo, D.; Russo, D.; Libonati, R.; Faraone, I.; Candido, V.; Picuno, P.; Andrade, P.; Valentao, P.; Milella, L. Influence of shading treatment on yield, morphological traits and phenolic profile of sweet basil (*Ocimum basilicum* L.). *Sci. Hortic.* **2019**, *254*, 91–98. [CrossRef]

23. Valera, M.D.; Molina, A.F.; Alvarez, M.A. *Protocolo de Auditoría Energética en Invernaderos. Auditoría Energética de un Invernadero Para Cultivo de flor Cortada en Mendigorría*; Instituto para la diversificacion y ahorro de la energia: Madrid, Spain, 2008.

24. Rasheed, A.; Na, W.H.; Lee, J.W.; Kim, H.T.; Lee, H.W.J.E. Optimization of greenhouse thermal screens for maximized energy conservation. *Energies* **2019**, *12*, 3592. [CrossRef]

25. Rafiq, A.; Na, W.H.; Rasheed, A.; Kim, I.I.T.; Lee, H.W.J.P.H.; Factory, P. Determination of thermal radiation emissivity and absorptivity of thermal screens for greenhouse. *Prot. Hortic. Plant Fact.* **2019**, *28*, 311–321. [CrossRef]

26. Ahamed, M.S.; Guo, H.; Tanino, K. Development of a thermal model for simulation of supplemental heating requirements in chinese-style solar greenhouses. *Comput. Electron. Agric.* **2018**, *150*, 235–244. [CrossRef]

27. Vanthoor, B.H.E.; Stanghellini, C.; van Henten, E.J.; de Visser, P.H.B. A methodology for model-based greenhouse design: Part 1, a greenhouse climate model for a broad range of designs and climates. *Biosyst. Eng.* **2011**, *110*, 363–377. [CrossRef]

28. Geoola, F.; Kashti, Y.; Levi, A.; Brickman, R. A study of the overall heat transfer coefficient of greenhouse cladding materials with thermal screens using the hot box method. *Polym. Test.* **2009**, *28*, 470–474. [CrossRef]

29. Geoola, F.; Kashti, Y.; Teitel, M.; Levi, A.; Brickman, R.; Esquira, I. *A Study of u Value of Greenhouse Films with Thermal Screens Using the Hot Box Method*; International Society for Horticultural Science (ISHS): Leuvenz, Belgium, 2011; pp. 367–372.

30. Rasheed, A.; Lee, J.W.; Lee, H.W. Evaluation of overall heat transfer coefficient of different greenhouse thermal screens using building energy simulation. *Prot. Hortic. Plant Fact.* **2018**, *27*, 294–301. [CrossRef]

31. Taki, M.; Ajabshirchi, Y.; Ranjbar, S.F.; Rohani, A.; Matloobi, M. Modeling and experimental validation of heat transfer and energy consumption in an innovative greenhouse structure. *Inf. Process. Agric.* **2016**, *3*, 157–174. [CrossRef]

32. Ahemd, H.A.; Al-Faraj, A.A.; Abdel-Ghany, A.M. Shading greenhouses to improve the microclimate, energy and water saving in hot regions: A review. *Sci. Hortic.* **2016**, *201*, 36–45. [CrossRef]

33. Abdel-Ghany, A.; Al-Helal, I.; Alkoaik, F.; Alsadon, A.; Shady, M.; Ibrahim, A. Predicting the cooling potential of different shading methods for greenhouses in arid regions. *Energies* **2019**, *12*, 4716. [CrossRef]

34. Lee, M.; Lee, I.-B.; Ha, T.-H.; Kim, R.-W.; Yeo, U.-H.; Lee, S.-Y.; Park, G.; Kim, J.-G. Estimation on heating and cooling loads for a multi-span greenhouse and performance analysis of pv system using building energy simulation. *Prot. Hortic. Plant Fact.* **2017**, *26*, 258–267. [CrossRef]

35. Hassanien, R.H.E.; Li, M.; Lin, W.D. Advanced applications of solar energy in agricultural greenhouses. *Renew. Sustain. Energy Rev.* **2016**, *54*, 989–1001. [CrossRef]

36. Ahamed, M.S.; Guo, H.; Tanino, K. Sensitivity analysis of csgheat model for estimation of heating consumption in a chinese-style solar greenhouse. *Comput. Electron. Agric.* **2018**, *154*, 99–111. [CrossRef]

37. Wei, B.; Guo, S.; Wang, J.; Li, J.; Wang, J.; Zhang, J.; Qian, C.; Sun, J. Thermal performance of single span greenhouses with removable back walls. *Biosyst. Eng.* **2016**, *141*, 48–57. [CrossRef]

Article

A SWOT Analysis for Offshore Wind Energy Assessment Using Remote-Sensing Potential

Meysam Majidi Nezhad [1], Riyaaz Uddien Shaik [2], Azim Heydari [1], Armin Razmjoo [3], Niyazi Arslan [4] and Davide Astiaso Garcia [5,*]

[1] Department of Astronautics, Electrical and Energy Engineering (DIAEE), Sapienza University of Rome, 00184 Roma, Italy; meysam.majidinezhad@uniroma1.it (M.M.N.); azim.heydari@uniroma1.it (A.H.)

[2] School of Aerospace Engineering, Sapienza University of Rome, 00184 Rome, Italy; riyaaz.shaik@uniroma1.it

[3] Escola Tècnica Superior d'Enginyeria Industrial de Barcelona (ETSEIB), Universitat Politècnica de Catalunya (UPC), 08028 Barcelona, Spain; arminupc1983@gmail.com

[4] Department of Geomatics Engineering, Cukurova University, Ceyhan Campus, 01950 Ceyhan, Turkey; niyazi.arslan@gmail.com

[5] Department of Planning, Design, and Technology of Architecture (DPDTA), Sapienza University of Rome, 00197 Rome, Italy

* Correspondence: davide.astiasogarcia@uniroma1.it; Tel.: +39-06-4991-9174

Received: 13 August 2020; Accepted: 12 September 2020; Published: 14 September 2020

Abstract: The elaboration of a methodology for accurately assessing the potentialities of blue renewable energy sources is a key challenge among the current energy sustainability strategies all over the world. Consequentially, many researchers are currently working to improve the accuracy of marine renewable assessment methods. Nowadays, remote sensing (RSs) satellites are used to observe the environment in many fields and applications. These could also be used to identify regions of interest for future energy converter installations and to accurately identify areas with interesting potentials. Therefore, researchers can dramatically reduce the possibility of significant error. In this paper, a comprehensive SWOT (strengths, weaknesses, opportunities and threats) analysis is elaborated to assess RS satellite potentialities for offshore wind (OW) estimation. Sicily and Sardinia—the two biggest Italian islands with the highest potential for offshore wind energy generation—were selected as pilot areas. Since there is a lack of measuring instruments, such as cup anemometers and buoys in these areas (mainly due to their high economic costs), an accurate analysis was carried out to assess the marine energy potential from offshore wind. Since there are only limited options for further expanding the measurement over large areas, the use of satellites makes it easier to overcome this limitation. Undoubtedly, with the advent of new technologies for measuring renewable energy sources (RESs), there could be a significant energy transition in this area that requires a proper orientation of plans to examine the factors influencing these new technologies that can negatively affect most of the available potential. Satellite technology for identifying suitable areas of wind power plants could be a powerful tool that is constantly increasing in its applications but requires good planning to apply it in various projects. Proper planning is only possible with a better understanding of satellite capabilities and different methods for measuring available wind resources. To this end, a better understanding in interdisciplinary fields with the exchange of updated information between different sectors of development, such as universities and companies, will be most effective. In this context, by reviewing the available satellite technologies, the ability of this tool to measure the marine renewable energies (MREs) sector in large and small areas is considered. Secondly, an attempt is made to identify the strengths and weaknesses of using these types of tools and techniques that can help in various projects. Lastly, specific scenarios related to the application of such systems in existing and new developments are reviewed and discussed.

Keywords: marine renewables; remote sensing; offshore wind; SWOT(strengths, weaknesses, opportunities and threats) analysis

---

1. Introduction

Human societies are currently facing global warming, and national governments are working to find solutions against the climate crisis, by promoting the installation of renewable energy sources (RESs) to replace fossil fuels. This energy transition could avoid the emissions of GHG (greenhouse gases) [1,2] and reduce the use of carbon fuel reserves [3]—minimizing at the same time geopolitical conflicts to access oil and gas sources [4]. Additionally, this green energy transition is creating new challenges in different sectors and consequentially new job opportunities in economic, technological, and environmental topics [5], together with other kinds of economic and social benefits [6]. With the aim of reducing carbon exploitation, many power energy converters have been designed and installed, considering the specific contexts of each analyzed area. Depending on the sources of power extracted from seas and oceans, two different categories should be defined: "blue energy" (BE) and "marine renewable energies" (MREs). Blue energy, such as salinity gradients, thermal differences and MREs including sea surface waves, sea current, tides, wind, geothermal and solar. Assessing the renewable energy sources (RESs) availability is important in developing short and long-term planning [7,8]. In this regard, wind energy could be one of the safest and most reliable sources of renewable energy [9]. To use these renewable sources, many aspects must be examined. Among these, it is fundamental to assess the exact amount of power for each type of energy converter. This has given more attention to the development of new offshore solutions, such as wind turbines with larger rotors, deep water foundation and floating platforms [10]. Northern and central European countries (ECs) have a long history of designing, developing and installing offshore wind farms [11], since the installation of first prototypes of bottom fixed and floating offshore wind farm in Baltic Sea and Scotland pilot park [12]. Nearly 90% of the world's MREs are in Europe. However, the proportion of the Mediterranean in the use of this vast resource is extremely low [13].

Among those with a feasible amount of wind energy source, the best suitable sites for offshore wind should be selected mainly according to the optimal combination of water depth and distance to the shore [14]. Water depth is a key factor in better understanding the dynamics of the marine environment, in predicting tides, currents and waves and planning offshore facilities and infrastructure such as wind turbines. Selecting suitable places with the optimal combination of water depth and distance to the beach is a complex task and should be carefully examined. The coefficient of water depth in the desired area is the basic parameter in the type of wind farm installation, mainly due to the maintenance and installation cost increase. In addition, the distance from the desired point to the power grid is very important, which increases due to the greater distance, which leads to more use of cables and batteries. Floating platforms could help in this regard which is the current trend to move considering deeper waters. Furthermore, there are more crucial factors/limitations influencing offshore wind (OW) applications, such as ships sea routes for marine transportation, migrating birds, economic activities (e.g., fisheries areas), environmental constraints (marine protected areas) and landscaping view. To combine economic sustainability and public acceptance, the concept of floating marine turbines operating away from coastal areas has also begun and expanding in the Mediterranean Sea [14]. Together with the growth of onshore wind farms installations, there is an expectation for significantly increasing the new wind farms in offshore marine areas, mainly thanks to floating platform technologies [15]. The dramatic growth of new technologies has led to an immediate revolution in the use of new offshore wind farms, statistically, with the wind farm sector in continental Europe showing an annual growth of 101% in 2017 [16].

In general, the installation capacity of wind farms in 2019 for European countries (ECs) is 21.1 GW and there is an expectation that in 2020, total OW energy production could reach 25 GWh, and by 2030, it could reach to 70 GW in offshore installation capacity [17,18]. Totally, ECs have 4149 grid-connected wind turbines and 81 offshore wind farms installed, which are used only in 10 countries of northern Europe. According to data from 2017, about 50% of offshore wind farms in continental Europe were installed in UK (which is 53% of the net 3.15 GWh of installed capacity in Europe). By 2024, Europe's total installed capacity is expected to reach 29.8 GW, expanding at an annual growth rate of 12% [19]. On the other hand, despite all these considerations, there are still no OW farms installed in the Mediterranean Sea, mainly for the water depth that usually does not allow the installation of bottom fixed wind turbines; anyway there are attractive hot spots for future developments of OW in the Mediterranean [20].

The first step in installing wind farms is to evaluate wind sources in focused areas or hot spots. Traditional methods, such as cub anemometers, are frequently used to measure wind sources, by installing calibrated calipers on tall masts. It should be noted that the height of the wind gauges is directly related to the height of the installed wind turbines. Due to the significant growth of technology, the height of wind turbines is constantly increasing. Consequentially, taller masts are needed, increasing the installation costs and the operation and maintenance efforts. In addition to all the above consideration, it should be noted that natural and human obstacles are the most important factors for the installation of anemometers. The fewer natural and human factors in the area, the fewer wind gauges are needed. However, if there are natural and human obstacles such as cities, more wind gauges are needed [21]. On the other hand, due to the significant development of on-site remote sensing sector, LiDAR (light detection and ranging) and SODAR (sonic detection and ranging) tools could be applied. Goit et al. [22], explained the though reconstruction from LiDAR-measured radial wind speed to wind speed vector is a challenge, LiDAR-based wind speed measurements are undergoing a significant increase in interest for wind energy application. Here, the study employed the scanning of Doppler LiDAR for assessment and comparison. First, the evaluation of the effect of carrier-to-noise-ratio (CNR) and data available on the quality of scanning LiDAR measurements was done. Then, it was proposed to reconstruct the wind fields from plan-position indicator (PPI) and range height indicator (RHI) scans of LiDAR-measured line of sight velocities. It was observed that an internal boundary layer with strong shear could be developed from the coastline. Lastly, PPI scan was involved to measure the flow field around a wind turbine and validate wake models. Chaurasiya et al. [23] investigated how to increase the confidence of RS technique to compute Weibull parameters at higher heights for assessment of wind energy resource. It is known that RS techniques are gaining attention worldwide for their comprehensive assessment of wind source in flat, complex, and mountainous terrain. The 10 min average time series wind speed data for the period of one month (in September 2014) were recorded simultaneously at 80 m and 100 m using the cup anemometer installed in the proximity of a 120 m meteorological mast, second wind triton SODAR (sound detection and ranging) and continuous wave wind LiDAR (light detection and ranging). Nine different methods have been implemented to analyze and obtain Weibull parameters on the data obtained from the measurements. Totally, there is an expectation that the outcome of this study could encourage the deployment of remote sensing techniques.

However, this equipment is very expensive and needs to be installed in a study area for more than one year to get enough data; on the other hand, high maintenance and repair costs are required [24,25]. Consequentially, it is very important to develop new methods that can help to identify suitable areas faster and economically. Satellites are the only tool that can measure the average, minimum and maximum wind speed in a study area (hot spot areas) in the shortest possible time. It should be noted that the reasons for the popularity of these data in the research and academic communities can be mentioned as follows: (i) This data are generally available for free (open access), (ii) They can cover a period of more than 40 years. Due to the fact that it is not possible to install ground wind gauges (such as, cub anemometers) in different areas, due to its high cost, satellites are the only device that can cover areas for more than a year, which is an important factor in assessing wind resources [21]. Satellite technologies for observing, reporting, and evaluating RES potential have led to a revolution in the installation of energy converters in new locations that have not previously been considered. In addition, due to the increase in surveys to identify industrial wind at an altitude of more than 100 m, various atlases were generated as an outcome. It takes long time to install anemometers on-site to measure industrial winds at higher altitudes which can be done with satellite data. However, to bring out the best options and strategies, all aspects of SWOT (strengths, weaknesses, opportunities and strengths, weaknesses, opportunities and threats) must be considered. However, all aspects of SWOT need to be considered to point out the best options and strategies [26]. A SWOT analysis can be used to achieve this goal. This type of analysis, derived from an interdisciplinary approach, can identify barriers and factors influencing the development of marine renewables.

Pisacane et al. [27] explained the current prospects for the exploitation of power plants in the Mediterranean Sea, outlining and discussing challenges, opportunities, and limitations for the deployment of power converters. It was stated that blue energy conversion technologies are now ready to be fully deployed in the device farms. Goffetti et al. [26] described a SWOT analysis of strategic plans for marine renewable energy technologies to minimize and maximize inefficiencies and energy production. SWOT analyses have been used for the navigation of renewable energy technologies and identifying key or hot spots points in various sectors including social, economic, legal, technological, and environmental. Nikolaidis et al. [28] investigated the status of marine renewable energy potential in the Mediterranean Sea and especially around the island of Cyprus. According to their study, OW energy in the Mediterranean Sea is the prominent outlet followed by marine biomasses. On the other hand, they explained that the main physical parameter for developing marine renewable energy projects around that islands is bathymetry. Azzellino et al. [29], using a spatial planning approach, explained the feasibility of installation by choosing a best location for wind turbine generators (WTGs) and wave energy converters (WECs). They proposed a quantitative spatial approach to identify potential sites of interest for the development of marine renewables with an effective perspective, by considering and minimizing potential environmental impacts. The obtained results showed that the Tyrrhenian coastline surrounding the island of Elba, the Northern and Western Sardinian coasts and the Adriatic Sea and Ionian coastal waters, were the most suitable sites for installing marine energy converters. Moreover, further studies about SWOT analysis locations are available.

In the light of the above considerations, the main aim of this research is to develop for the first time a framework on SWOT analysis for investigating the strengths, weaknesses, opportunities and threat of remote sensing (RS) techniques to measure potential power from OW installations. In particular, the present study aimed at elaborating a SWOT analysis for assessing the wind energy potential with RS techniques in the biggest islands of the Mediterranean Sea: Sicily and Sardinia. This study raises a better understanding of how to use RS technology to replace traditional wind measurement tools. The results of the SWOT analysis are expected: (i) to highlight satellites' ability to measure marine renewables; (ii) to identify pros and cons of using these techniques.

1.1. Synthetic Aperture Radar (SAR) Satellites

The first European Space Agency (ESA) satellite was launched in July 1991 and called the European remote sensing satellite (ERS-1) [30], which is a C band (5.3 GHz) microwave instrumentation (AMI) satellite [31]. The ESA has recently designed, developed, and launched new satellites so-called Sentinel family providing free data after a simple registration. These satellites could be used to conduct research on various parameters of RESs, for example wind speed, wind direction, wave height, tidal, thermal ocean water and ocean depth measurement. Several SAR satellites and scatterometer, such as QuikSCAT, OSCAT, ASCAT-A, ASCAT-B and Sentinel have already in use for this purpose. Sentinel family which is a group of satellites orbiting around the earth with varying revisiting time for observing land, ocean and atmosphere from space and then for providing us the data free of charge anytime around the year (24/7 and 365 days). One of the most ambitious and world's largest earth observation program in existence today is the Copernicus. Previously known as GMES (global monitoring for environment and security), aiming to tackle environmental challenges with a fleet of autonomous satellites. Starting from global warming to land use change and the atmosphere, Copernicus gather earth data from space and in the center of it all is its Sentinel family of satellites. Satellites are largely used for different kind of applications [32].

Karagali et al. [33], explained the characterization of the near-surface winds over the northern European shelf seas using satellite data, including the inter- and intra- annual variability for resource assessment purposes. Comparison of mean winds from QuikSCAT with reanalysis fields from the WRF (weather research and forecasting) model and in situ data from FINO-1 offshore research mast was carried out. By this study, the applicability of satellite observations as the means to provide useful information for selecting areas performing higher resolution model runs or for mast installations. It is observed that there were biases ranging mostly between 0.6 and -0.6 ms^{-1} with a standard deviation of 1.8–2.8 ms^{-1}. The combined analyses of inter- and intra- annual indices and the wind speed and direction distributions allow the identification of three subdomains with similar intraannual variability. High-resolution satellite SAR wind fields were depicted to observe the local characteristics from the long-term QuikSCAT wind rose distributions. The WRF reanalysis dataset misses seasonal features observed for QuikSCAT and at FINO-1 winds.

Bentamy et al. [34] considered some more specific areas for studying and assessing the offshore wind power potential. To achieve their objective, requirement was given on wind speed and direction with enough spatial and temporal sampling under all weather conditions during day and night. For more than 12 years of remotely sensed consistent data that were retrieved from ASCAT and QuikSCAT scatterometer estimations, were used to find conventional moments associated with wind distribution parameters and then the latter comparable to wind observations from meteorological stations. Further improvisation was carried out by combining in situ and scatterometer wind information. Wind statistical results were used to study the spatial and temporal patterns of the wind power. Also, they depicted the main parameters characterizing wind power potential such as variability, mean, maximum energy, wind speed and intraannual exhibit seasonal features and interannual variability and then found the differences between the wind power estimated for northern and for southern Brittany. Signell et al. [35] detailed spatial structure of jets using the in-situ observations that were carried out on northeast Bora wind events in the Adriatic Sea during the winter. For this, high resolution spaceborne RADARSAT 1 SAR images collected during the active Bora period of the year 2003, created a series of high-resolution maps of 300 m dimension. Along with the previous observations on Bora winds, in this study, it was understood that along the Italian coast, several images show a wide (20–30 km) band of northwesterly winds that abruptly change to northeasterly Bora winds further the offshore. It was concluded by meteorological model that northwesterly winds are consistent with those of a barrier jet forming along Italian Apennine mountain chain.

1.2. ECMWF Reanalysis

ECMWF is the European Centre for Medium-Range Weather Forecasts, this organization is advancing numerical weather prediction through global collaboration. The ECMWF reanalysis dataset was chosen for different studies, because of its high temporal and spatial resolution [36]. ECMWF dataset covers a vast timespan starting from 1979 to the present time [37]. ECMWF Reanalysis datasets, including ERA5, ERA-interim, ERA-interim/land, CERA-SAT, CERA-20C, ERA-20CM, ERA-20C. Furthermore, ECMWF data gathers information from the global observation system comprehending different kind of satellites, meteorological, buoys, weather stations and ships.

These limitations are mentioned above, that do not include RS satellite technology and reanalysis datasets [38], such as ECMWF, MERRA, NARR and ERA. A reanalysis dataset is the only source resolved long-term information of spatial wind information at wind turbine height. It provides data and potential assessment of this sector strongly related to the quality form of the meteorological situation [39]. Due to high costs and limitations of the measuring device, including coastal stations, buoys, ships, masts, the use of this kind of dataset can give possibility to understand wind speed with good quality in wind farm site assessment. This step is a very important criterion in possible alternatives that can be provided by high-resolution reanalysis data. Furthermore, this kind of analysis can reduce the cost of on-site wind measurement [39]. Arun Kumar et al. [40], described the limits of offshore buoys, pinpointing that wind source assessment would be a challenging task. One of the best alternatives for inexpensive data and for filling the data gaps by providing a huge volume of data for extended periods is satellite RS. The assessment of wind source from single scatterometer could lead to inconsistency where there is a requirement of multiple satellite scatterometer. Therefore, four scatterometers viz. OSCAT, ASCAT-A, ASCAT-B and QuikSCAT with long-term in situ wind datasets over the North Indian Ocean were considered. It has been observed that QuikSCAT and OSCAT wind data have lesser bias with the range of 0.15 m/s (2.4%) to 0.83 m/s (15.1%) before adjustments. Linear regression was used for adjustment and the synergetic approach of linear regression adjustments and the combination of scatterometer data have resulted in smaller differences.

2. Case Studies

Many islands are not connected to the main grid and still significantly depend on fossil fuels for covering their energy needs. This approach is no more sustainable considering the economic aspects, environmental externalities, high costs in electricity generation and the high amounts of pollutant emissions. Furthermore, such isolation often leads to a high dependency on foreign countries, that could be solved with better solutions for increasing the renewables penetration in small or big islands [41]. Statistically, considering more than 85,000 islands all around the world, approximately 13% of them are inhabited with a population of around 740 million [42]. About 21 million people live on 2050 small islands, each with a population between 1000 and 100,000 inhabitants. Electricity demand in these islands is around 52,690 GWh; anyway, almost half of these islands is in the Pacific Ocean and still have no access to electricity [43].

As regard the Mediterranean Sea, satellite data can be used to measure renewable energy potential for exploiting marine renewables to meet islands energy needs. The exploitation of marine RESs in the Italian islands is crucial to gradually replace their heavy dependence on fossil fuels. On the other hand, Italy has a coastline of more than 8000 km, including 458 small islands with an interesting potential for installing WTGs and WECs [26]. Many of these islands have a higher wind potential compared to coastal areas in the mainland [44]. The highest wave energy potential in Italy is mainly located on the west coast of Sardinia and Sicily [45]; in particular, wave power at the coast is estimated as 10 kW/m in the west coast of Sardinia and 4.5 kW/m in the West Sicilian coastline [46]. Figure 1 shows the location of the two case study areas in the Mediterranean Sea. The red area (West of Sicily) covers three small Italian islands (Favignana, Marettimo and Levanzo) and the green area covers the Southwestern region of Sardinia in same ERA-interim pixel size (6 pixels for each regions).

One important parameter for installing marine energy converters is bathymetry, which is difficult in the Mediterranean Sea and does not allow for the installation up to several hundred meters off the coast due to the high depths [47]. Bathymetry and power potential availability are key parameters to be considered for developing the technical feasibility of designing and installing WTGs and WECs at sea [48]. However, technical, and economic feasibilities related to the use of floating platforms for wave and wind offshore farms are quickly improving.

In this context, wind is considered a very promising source for WTG installation and future WEC installation on the islands, respectively. Hence, various studies were carried out to identify the best places to utilize those energy sources. Even if each RES has limitations on the use and installation of converter devices, they can also use in combination. This combination of energies could be especially applicable to small islands in the Mediterranean, as they have many commonalities such as size, available RESs, weather conditions, population, and environmental constraints [41]. On the other hand, the two most important parameters to consider while identifying the feasible locations are tourist activity and landscaping constraints [49] because tourism is one of the key economic activities [50] in most of the Mediterranean Sea islands, mainly in summertime. The Mediterranean region is one of the most popular tourist destinations in the world and attracts one-third of international tourists.

The number of these tourists is expected to reach 500 million in 2030 [51]. The increase in significant energy demand during the summer season can be covered by these potential renewable energy sources.

Furthermore, many of these islands are included in marine protected areas, where the installation of power converters such as WTGs and WECs can be forbidden for environmental reasons. Generally, many factors and parameters must be specifically considered when suitable areas for wind turbine installation are close to environmental protected areas [52]. With respect to environmental issues, comparatively, fossil fuel power plants pollute more than wind farms. For example, noise caused by rotor blades, could affect the behavior of living species such as dolphins, fish, bats and birds [53]. Industry and researchers are working to reduce the negative effects of WTGs on wildlife by taking preventive measures [6] by choosing the proper location of a wind farm to reduce the bird mortality rate. Many studies were aimed to assess and mitigate environmental impacts of WTGs in marine areas, such as birds and bats and other wildlife species [54].

The installation of OW farm technologies dramatically changes the shape of the offshore realm, which may cause conflicts between communities and developers. Considering this issue, evaluation and analysis of wind sources play a key role in selecting suitable locations for the construction of OW farms. Wind energy assessment, planning and installations [55] must be carried out considering important parameters such as mean wind speed (MWS), wind energy density (WPD) and Weibull parameters [56].

Figure 1. Two case studies near the Sicilian and Sardinia islands (with red and green boxes) in the Mediterranean Sea.

3. Material and Methods

3.1. Satellites and Reanalysis Analysis

The developed method is based on an integrated approach for the preliminary wind speed assessment using Sentinel 1 and reanalysis data by, (i) Sentinel application platform (SNAP) software and (ii) environment for visualizing images (ENVI) software [57]; (i) SNAP stands for sentinel application platform, which is a common architecture for all Sentinel toolboxes, was jointly developed by Brockmann Consulting, SkyWatch and the C-S. This is an ideal software for Earth observation (EO) processing and analysis due to the various technological innovations such as modular rich client platform, extensibility, portability, tiled memory management, generic EO data abstraction and a graph processing framework [58].

(iii) ENVI Software, meaningful information from imagery can be extracted from satellite imagery using this software to make better decisions [59]. This is one of the popular and user-friendly software in the field of RS, which is mostly used by RS scientists, image analysts and geographical information system (GIS) professionals. This software could be accessed from the desktop, in the cloud and on mobile devices and could also be customized through an API to meet specific project requirements. It uses scientifically demonstrated analytics to deliver expert-level results and also various businesses and organizations preferred ENVI because it has shown easy integration with existing workflows, supported most popular sensors and could easily be customized to meet unique project requirements [59]. ROI (region of interest) tool in ENVI is one of the most used tools in the many applications and have been in from many years and in many processes since the development of its first version called ENVI classic. This tool is used to select the ROI in the satellite image for further analysis or vice versa. Usually, ROI can be selected with geometric shapes like square, polygon, etc., but the drawback of using ROIs is that they are based on image coordinates (number of rows and columns) rather than map coordinates which means they are not easily transferred between images of different sizes or projections. Map coordinate-based vectors (shapefiles and ENVI evf) are more frequently utilized because they are more portable between images and between image processing packages [60]. However, there are still many uses for ROIs and in ENVI there is a new method for their creation too. In this study, preliminary data were obtained by ECMWF reanalysis dataset. For ERA-interim processing, it was carried out in two main steps, namely: (i) analyze era-interim data with GIS software for mapping; (ii) wind speed analysis using the ROI tool.

The first step involves the use of GIS software for mapping wind potential and other parameters [61] in the study area as shown in Figures 2–5. The second step enables the user to analyze the potential of wind energy in a specific area focusing on different zones [62] or hot spots, such as the west part of Sicilian and Sardinia Islands. The ROI tool was used for classification, masking, and operations, also for automatically retrieving information and statistics about a specific area in a larger or smaller area. At this point, after identifying the specific area, all the layers can be merged as one layer to make a time series analysis according to a different time steps and research grope targets (per day, monthly or yearly).

Figures 2–5 show MWS (m/s), mean wind power (MWP), significant wave height (SWH) (m) and sea current speed (SCS) (m/s), analyzed using ECMWF reanalysis dataset for the Mediterranean Sea. As already mentioned, the biggest problem in the design, development, and installation of MRE converters in the Mediterranean Sea is water depth, due to the steep slopes around the shorelines.

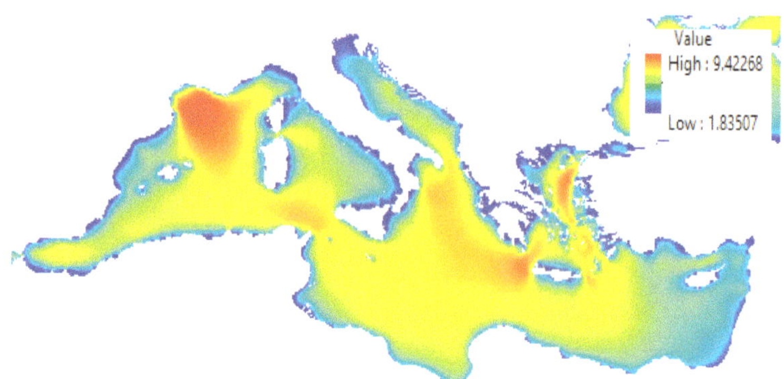

Figure 2. Mean wind speed (MWS) (m/s) with 10 m height in the Mediterranean Sea for the years from 2010 to 2017.

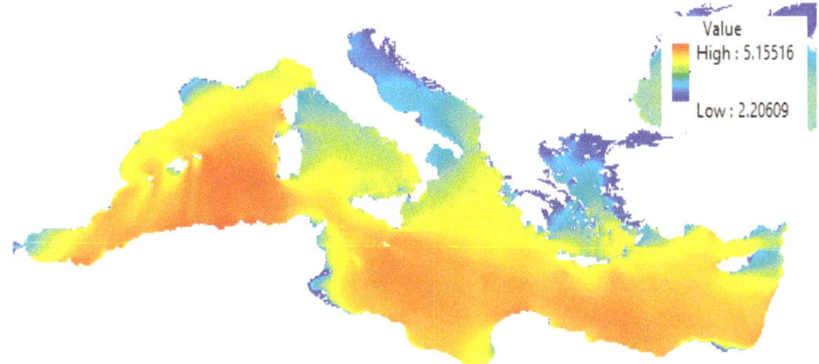

Figure 3. Mean wind power (MWP) (kW) with 10 m height in the Mediterranean Sea for the years from 2010 to 2017.

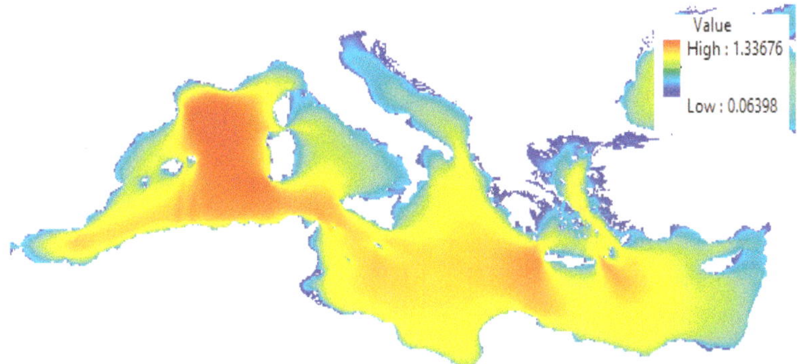

Figure 4. Significant wave height (SWH) (m) in the Mediterranean Sea averaged for the years from 2010 to 2017.

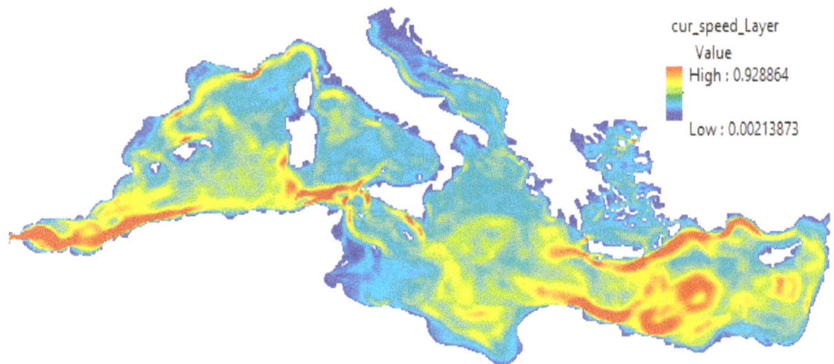

Figure 5. Sea current speed (SCS) (maximum speed over layer depth) averaged with 10 m height for the years from 2010 to 2015.

3.2. SWOT Analysis

A SWOT analysis was carried out to identify areas of interest for marine renewables installation around large or small islands. This type of analysis is used for small and medium strategic planning [27]. Moreover, some researchers suggested that SWOT analysis can be used as an appropriate tool for the development of sustainable energy strategy and strategic environmental assessment systems at national level [63].

The purpose of this type of analysis is to design a qualitative process structure that identifies the changes with main strengths, weaknesses, opportunities, and threats to the qualitative structure of the system under consideration. Currently, there are many research articles on the evaluation of renewable energy at different scales of data from satellite measurements and reanalysis source [64]. However, there is no shortage of such studies, although many authors aim to compare different sources of wind speed data and compare it with terrestrial ones. Overall, a SWOT analysis helps the identification of strengths and weaknesses of the strategy to achieve its goals, pinpointing opportunities to reduce the impact on the goals [27]. Tables 1 and 2 summarize a survey of the factors belonging to different SWOT compounds. Information was disaggregated into two subcategories to highlight specific topics for consideration and discussion. The next paragraphs are devoted to present sectorial specificities.

Table 1. SWOT (Strengths, Weaknesses, Opportunities and Threats) analysis of satellite potential.

Forces	Internal	External
Satellite Potential	Strengths ESA and Italian Space Agency (ISA). Open access and unlimited policy support. Wind atlas mapping (wind industry).	Opportunities Save time and money. Knowledge transfer from/to research centers and universities. A better understanding of forgotten areas of whole world (cover existing measurements gaps)
	Weaknesses The uniqueness of the available information due to the rotation of the satellite in the Earth's orbit.	Threats Specialized knowledge in various interdisciplinary fields.

Table 2. SWOT analysis of potential assessment aspects.

Forces	Internal	External
Potential Assessment	Strengths Marine large area observation. Find area with good geographic position and energy potentials (hot spot).	Opportunities Better understanding of sources. Reduce strategic risk. Maximum use of susceptible areas.
	Weaknesses Sources estimation in Italy is incomplete. Incomplete measurable parameters.	Threats Climate change.

These categories provide an overview of the factors for using wind speed measurement satellites in areas of interest with different perspectives which could help to better understand the capability of this tool in measuring wind energy. Furthermore, this analysis is based on a review of the literature in two categories in different dimensions and also uses the support tools used in the ODYSSEA project (the platform is available on, http://odysseaplatform.eu/).

4. Results

In this section, the results of the SWOT analysis were summarized, considering the two parts of RS potential and potential assessment.

4.1. Remote Sensing Techniques

Strengths: Recently, ESA designed, developed, and launched a new family of satellites called Sentinel (includes S1 to S6) as a part of the Copernicus Program. Sentinel-1 (S1, SAR) is able to perform very detailed analysis in this area and would like to thank the different polarization modes: single polarization (vertical-vertical (VV) or horizontal-horizontal (HH) and dual-polarization (VV + VH or

HH + HV) [65]. The VV polarization is very useful for detecting wind speed in ocean and also for understanding the different kind of ocean and sea activities such as fisheries, ship routing, coastal surveillance, offshore installations and exploration. The main reason to select the VV polarization is because of its success in detecting wind speed, since this kind of polarization is sensitive to sea and ocean surface roughness (sea surface water). Images obtained at VV polarization by the SAR satellite are highly sensitive to wind speed variations by means of RMSE which is lower for VV polarization than HH polarization for Sentinel-1B [66]. For C band SAR images analysis, such as Sentinel 1, the C-band model CMOD (C geophysical model function) family (such as CMOD 4, CMOD 5, CMOD 5.n, CMOD 7) and a new model function called C-SARMOD2 can be used. Measuring surface roughness caused by wind is an important feature of the SAR images. SAR satellite capability depends on Doppler information to achieve good resolution in the along-track direction [31].

The Italian Space Agency, along with the Italian Ministry of Defense have developed the COSMO-SkyMed system which is the unique constellation of four radar satellites for earth observation. Those four radar satellites of COSMO-SkyMed system have advanced technology and uses high-resolution radar sensors to observe the earth's day and night, regardless of weather conditions with varying revisit time. Main theme areas are emergency prevention, strategy, scientific and commercial purposes, providing data on a global scale to support a variety of applications among which forest & environment protection, risk management, natural resources exploration, land management, maritime surveillance, defense and security, food and agriculture management. The COSMO-SkyMed satellites have main payloads of X-band, multiresolution and multi-polarization imaging radar, with various resolutions (from 1 to 100 m) over a large access region. Since, it is equipped with a fixed antenna, having electronic steering capabilities that could manage many operative modes for the image acquisition and for internal calibrations. The nominal incidence angles varied between 20° and 59°.

SAR images have great potential for the observation, monitoring and detection of marine sources. It is important to have a wind parameters long-term reference analysis that can cover a large geographical scale. This is more important when many of the observations made from ocean tool measurements are scattered [67]. One of the most important reason to use S1 satellite data and software is because of its free access, supported by unlimited policy with just a sign-up before trying to download the images. However, the data are systematically provided by delivering within an hour of reception for near-real-time (NRT) emergency response, within three hours for NRT priority areas and within 24 h for systematically archived data. Images taken from satellites at different frequencies are used to analyze and map wind parameters at of the seawaters. Recently, the OW field retrieval method has been developed based on satellite data sources and image processing techniques [68]. In many of the Mediterranean coastal, nearshore, and offshore areas, there are no observations of the tall tower for validation. Hence, maritime validation relies on reports of ships and buoys.

Not only SAR satellite imagery can provide OW field data with a long time-series of large and small zones, but also satellite imagery is playing a significant role in offshore observation and research. Even though sea-level wind measurements can be carried out using buoys in different periods with very high resolution, these buoys are usually installed at a distance of 10 to 100 km from the shoreline indicating the possibility of the lack of access at greater distances [69]. Ocean winds recorded from scattering and radiometers have a higher temporal resolution [70]. This higher temporal resolution can be improved by using the cooperation of several satellite data. Due to the increase in the number of observations and time resolution, the accuracy of OW sources estimation can be gradually improved [71].

Furthermore, the ERA-interim reanalysis dataset is a reanalysis project dataset designed and developed by the ECMWF [37]. The ECMWF uses predicted models and data capture techniques, including 4D analysis with a 12 h analysis to describe the atmospheric parameters of the land and oceans such as wind speed, evaporation, surface pressure, surface roughness and surface net solar radiation [72]. In ECMWF dataset, the wind speed at 10 m height, 10 m U wind component, 10 m wind speed and 10 m V wind component are available since 1979. The intended spatial resolution is characterized by monthly, yearly and four hourly intervals for each day [73].

Even though the vastness and infinity of wind energy in many parts of the world are having good wind power [74], the problems related to the determination of wind measurement accurately making it out the scheme more difficult. The availability of meteorological and floating data information especially in coastal areas, making the situation difficult by lack of access to the complete data set for a study area [75]. On the other hand, good data are available to identify suitable locations through meteorological models and satellite observations. It should be noted that the reanalysis data collects a complete form of data on terrestrial existence, for example, meteorological stations, buoys and cub anemometers, ships and satellite data, which can provide a more accurate display of wind resources on a scale. Such data are regularly monitored with high quality without delay (unlike floating devices: buoys and cub anemometers). In this case, reanalysis data showed the lowest overall error compared with buoys and ship data [76].

Due to the dramatic increase in the use of renewable energy around the world, the need to identify suitable areas for installation and the size of wind farms has increased. In this case, the reanalysis data can be used with a great ability to identify these areas. For example, at different altitudes, it can measure the wind for industrial use of the area, so we called industrial wind. Which can show itself in the view of an accurate wind atlas. Wind atlas contains different wind parameters in different regions, for example, maps, wind speed and direction, time series and frequency distributions. A wind atlas covers the average of important wind parameters at different altitudes for long periods that can come to governments, companies, and academic projects, for example (https://globalwindatlas.info/). An atlas map of offshore wind energy potential that considers all wind parameters can identify potential areas for the nearshore and offshore wind farms installation for later stages [77].

Opportunities: RS data and techniques guarantee a high level of reproducibility because S1 data (SAR) is free and has global coverage. To be more precise and accurate, SAR images with longer time intervals should be used. SAR satellites have several advantages, such as the high spatial resolution, the coverage of large areas, obtained in all-weather conditions, day and night (24 h) [78]. The ability to identify hot spots or focus on ROI makes it an interesting tool for preliminary analysis with different goals and by different target groups. On the other hand, the researcher can obtain data over a long period of time. Many researchers used the reanalysis dataset, which is long term time series of data on wind speed analysis, for network integration studies on wind potential areas [79]. The most important advantage of reanalysis dataset is that they are generally free. This type of information received from the global observation system are made up by different observations tools such as satellites, meteorological stations, and ships to cover a large area [80].

Weaknesses: There are several sensor options for measuring wind at sea and ocean waters using satellite RS, but those satellite and sensors like many other measurement tools, has limitations. For this reason, many researchers have used synthetic winds derived from multimeter scattering, radiometry, and reanalysis data [81,82]. In addition, mostly the assessments were confined to a single or dual scatterometer or using different data sources like reanalysis model, a single scatterometer data is limited to a specific period of time [81]. On one hand, SAR satellite images relative with ocean and sea water surface usually manifests expressions of atmospheric phenomena occurring in the boundary layer can be attributed to the phenomenon [83], such as boundary layer rolls, atmospheric convective cells, atmospheric internal gravity waves, tropical rain cells [84]. Furthermore, SAR satellites have a limit on scanning different areas, which can include one or two scans per week and/or day and are not universal. Much of these satellite data can be accessible with some restrictions of user need (users need a proposal), such as TerraSAR-X, COSMO SkyMed and Radarsat-2. These restrictions also include time constraints with the launching time of satellites. The accuracy of offshore wind assessment using SAR satellite imageries may be affected by different reasons, such as land contamination and lower water depth in the coastline areas [56].

Other factors such as image quality, in-orbit time of the satellite and also hard targets such as oil spills, land, islands, ships, WTGs and WECs are limiting ability to measure wind at the ocean and sea areas from the satellite's imageries. Especially while researchers using the SNAP software, all these hard targets should be masked at the first stage of the processing. For example, in SAR image processing for wind field estimation, the studied areas should be checked out for oil spill and removed from the image because of decreasing the water roughness of the sea surface [85]. Another important weaknesses, the wind direction (SAR images can be analyzed in SNAP software to achieve wind direction with 180 degree ambiguity) obtained from the SAR satellites cannot be verified because the available data will not really show the wind field in the coastal and open seas. This is the main reason for selecting local data (in situ data) in many studies. Another limitation of wind speed estimation based on satellite data are that they measure wind speed data at 10 m above sea surface water. By using this type of data to obtain information about the wind sources at a height of 100 hub meters, it is necessary to implement theoretical models using surface roughness coefficient [76].

Threats: Interdisciplinary knowledge is needed to work with RS methods as this involves obtaining and analyzing satellite data. Then, researchers need to be familiar with various software applications to properly analyze satellite data. It is known that satellites are highly capable of observing the Earth, since they have applications in many different fields. Hence, studying the specific application mainly depends on the software knowledge of researchers.

4.2. Potential Assessment

Strengths: Long-term data collected from meteorological stations have generally been used to investigate the wind energy potential for an area. Given the limitations of measuring and installing marine buoys in offshore areas, much research needs to be conducted to identify the areas that are not considered. In this regard, satellite data are used in many aspects, for example, marine engineering, numerical model, oceanographic, wind speed and wave height [86]. In this area, many studies have already been conducted by various researchers on the estimation of marine wind energy on spatial scattering data greater than 12.5, 25 and 50 km resolutions [87,88]. This wind maps can be improved either by one-kilometer grid resolution [89] from Sentinel 1 data. Majidi Nezhad et al. [57] explained a new method for assessing, reporting, and mapping the wind energy potential of sea areas using Sentinel 1 imageries in the Sicilian island of the Mediterranean Sea. First, they identified the hot spots for wind turbine and wind farm installation in large marine areas and accordingly estimated the average wind potential in small areas around the islands. Sentinel 1 satellite images have been analyzed by using the SNAP software and then the wind parameters mapped in the GIS software. At the end of the SAR imagery analysis and mapping, the mean wind regime was extrapolated using the ROI tool and was used as an input data to train and test the proposed forecast model [68].

Figure 6 shows the mean wind speed (MWS) per (m/s) in two different cases (Sicilian and Sardinia islands) using ECMWF reanalysis dataset (between 1979 to 2019). Mediterranean Sea islands usually have higher average wind energy compared to the Mediterranean Sea region in the mainland (Figure 7). The case studies (Sicily, Sardinia) showed wind speeds are higher than the European continental shores. The main reason for this situation is the lower surface roughness (natural barriers) of the ocean surface compared to the ground land.

According to various studies conducted in the Mediterranean Sea, the highest values of wind speed can be observed in the Aegean Sea, the Gulf of Lyon and the Alboran Sea with wind speeds of more than eight meters per second. In the Mediterranean Sea, some areas such as the Aegean Sea, the Strait of Gibraltar, the Gulf of Lyon and the area between Sicily and the coast of Tunisia also have wind speeds of about seven meters per second. The highest increase in wind speed in these areas is observed in spring and summer. This also causes seasonal fluctuations in these regions. Furthermore, during the winter, a significant reduction in wind energy potential is observed in the central parts of the Mediterranean Sea, as in the coasts of Libya and Egypt [46].

These areas, given their good potential, can be called hot spot areas that can be focused on installing power converters. Majority of the Mediterranean islands are characterized by a strong economy from tourism, and these hot spots areas were used to generate renewable energy for the self-sufficiency of the islands also during the seasons when the energy demand significantly increases.

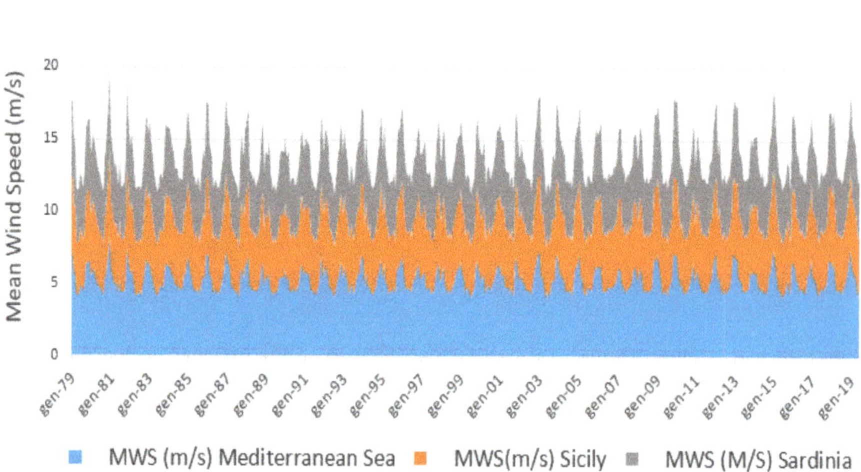

Figure 6. Mean wind speed (MWS) per (m/s) with the 10 m height in two different cases and Mediterranean Sea using European Centre for Medium-Range Weather Forecasts (ECMWF) reanalysis dataset (between 1979 to 2019).

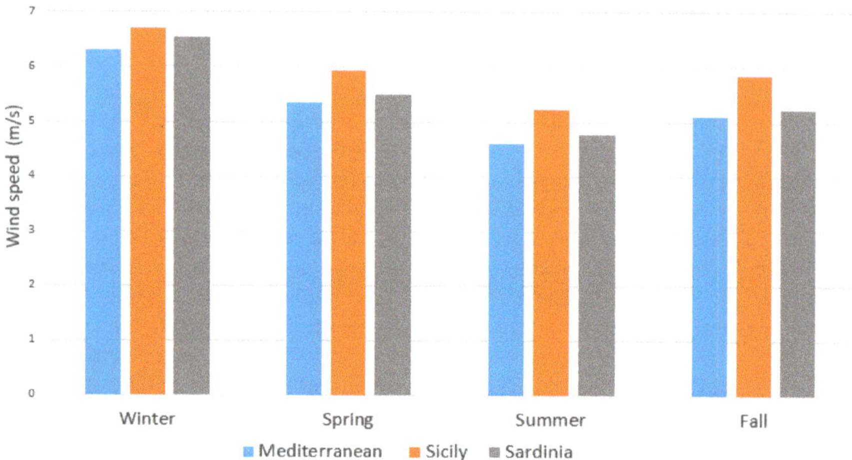

Figure 7. Seasonality wind speed (m/s) for case studies.

Opportunities: Since none of the wind farms were shut down to date, it can be treated as a new industry and also as a starting point, given its partnerships with the oil and gas industry [89]. One of the main benefits of OW energy are higher wind energy sources and potential electricity generation compared to land-based wind farms due to the higher wind speeds in the seas and oceans. It also has limited areas to explore given the greater scope for installing marine farms [90]. Sites with a short distance from the coastline are the most attractive location for wind farms installation, since installation costs increases with increase in the distance from the shore, mainly due to two related factors: water depth and cable cost [67]. Wind energy is the most popular type of renewable energy and its popularity is directly related to harmlessness. It is economical and environment friendly and can play an important role in reducing CO_2, SOx and NOx [10]. Using satellites to monitor wind energy and analyze changes in offshore wind energy is crucial for installing wind farms and for observing large areas. Such analysis often requires case study data that are particularly important for developing accurate business models for working on new offshore wind farms, considering that their number is expected to dramatically increase in the following years.

Weaknesses: The potential of Italy's OW energy source is outstanding in many places. The Italian government set a goal of 950 MW for OW by 2030 in its national energy and climate plan presented to the European Commission. Therefore, there are good strengths and opportunities for the Navy in Italy. However, the development of the Italian naval force is still at an early stage and still faces many challenges and threats. Government can set the strongest possible priorities for the future development of maritime zones in the best possible way with the clear and coherent maritime policies they adopt. Such policies should reflect specific goals aimed at establishing hydro power as a means of achieving some reduction in carbon dioxide and clean energy production [7]. Many offshore wind projects in Italy have been cancelled mainly due to lack of funding or opposition from local authorities and the small distances from shore. Moreover, many other projects are in a dormant state during the initial planning phase. OW farms continue to suffer from weaknesses, as prices within the OW power network were uncertain and so investment risk was higher. On the other hand, the development of OW power involved several parts such as planning conflicts between various departments and dealing with problems [91]. Large-scale wind turbines have become a mainstay of technology development in the world. Many countries declared that they were producing more wind turbines and the power plant was concentrated at eight megawatts to four megawatts [91].

Threats: Wind energy potential varies with the cube of wind speed, making minor changes to wind circulation patterns and severity will have a profound impact on future wind power generation. This indicates the sensitivity of wind energy which is related to climate change and future changes. Moreover, wind flow characteristics can also severely affect the ability of wind energy potential usage [92]. For example, a change in the year-to-year wind velocity impacts on electricity generation capability, and the higher the volatility, the more variable the power generation may be, which can cause problems in demand on the electricity grid. On the other hand, there is a decrease in the profitability of wind farms. Considering this parameter, this is an important issue for the economic feasibility of a wind farm having lifetime of about 20–25 years [92]. Climate change also affects other factors, such as water depth and distance at the coastal areas.

The high cost of investment is one of the most important reasons limiting the development of OW farms construction. OW energy sector includes many professional expertise and institutions: the maritime department, the environmental protection agency, the fisheries and military departments. The coordination between them could be very difficult, especially during the planning and approval processes [91].

5. Discussion

The purpose of this research is to identify and investigate the potential of satellite and reanalysis data for finding out the suitable areas for the installation of wind farms. A SWOT analysis was carried out to examine the various points of this study, i.e., remote sensing techniques and potential assessment which can help in better usages of new technologies such as RS for the development in the field of renewable energy by applying a suitable approach. Since there is an increasing attention in the widespread usage of clean technology in various societies, new measuring instruments need to be replaced by traditional instruments. For example, with increase in the height of wind turbines, the height of wind turbine towers also increases, which requires much money and time (which, in turn, takes more than one year to receive data from in situ measurements). With this approach, a preliminary study is required to find out the areas suitable for wind farms.

There is a possibility to use satellite technology for measuring various parameters mainly wind speed in different parts of the world. Today, the use of satellite data can be observed with a view of wind atlas, which is very suitable for the development of new wind farms. On the other hand, it helps governments, companies, and universities to focus more on forgotten areas for expanding the use of RESs. The factors influencing the development of renewable energy are not limited to their measurement technology, so factors such as social, economic, legal, technical, and environmental aspects must also be considered. In such cases, SWOT analysis would be helpful in selecting several factors and eliminating some other factors to limit research. In other point of view, it should not be forgotten that there is a possibility of ignoring some factor without the proper knowledge since this comes into other discipline of research. To avoid this problem, is important to work in collaboration with researchers from different fields.

Considering various aspects of the renewable energy assessment using satellite technology, the main issues to be considered are the obtainment of satellite data from space agencies or from the collaboration with universities with strong aerospace field. Moreover, the emergence of interdisciplinary disciplines and knowledge transfer between them is of importance. Efforts in increasing transparency, communication, and participatory policies may, in-turn, increase the use of each of these sectors in creating coherence between multiple stakeholders.

A better understanding of satellite measurement technology will help to use the potential of this tool in the economic field of projects in a better way, challenging the economic sector by reducing investment costs. By using satellite data, priority list can be created for the construction of new wind farms, taking factors into account such as environmental protected areas, tourist activity, maritime transport, fishing, etc. which could help in simplifying the decision of any of the desired sections.

Furthermore, an important issue to be discussed is that a wrong assessment of wind speed can lead to the failure of a wind farm project. Consequentially, due to high cost in using traditional methods, a preliminary study of the area is necessary. This initial study allows to obtain the rate of change of wind parameters in long- and short-term scenarios, starting from satellite data, facilitating their prediction in the future.

One of the major problems in most offshore areas is the lack of high-precision information. Due to the high cost of installing the device or the lack of calibration, these devices are sometimes less accurate, which, in turn, creates gaps in geographic and temporal data in the basin. In this regard, these gaps can be eliminated with the implementation of new methods using satellite data and a method to prioritize these gaps can be consequentially developed.

6. Conclusions

Trying to evaluate the technical suitability of areas for wind farms installation requires a clear conceptual framework that demands examining the potential of appropriate tools and installation locations. To achieve this goal, it is necessary to identify the technical potential that can be defined as a suitable framework for evaluating energy potential. The purpose of this article is to identify and study factors that could halt or encourage the development of satellite remote sensing with a focus on OW detection using a SWOT analysis, which is an appropriate tool for comprehensive research. The main limitation of SWOT analysis is its autonomy, which depends on the analyst's choice of factors to be considered.

In addition, another limitation of this type of analysis relates to the loss of information or compensation during processes when collecting information. Due to the development of satellite remote sensing techniques related to the wind field estimation from sea surface water, SWOT analysis is important to better-understand the importance of satellite data in offshore region. Obtained results pinpointed that there are currently several options for measuring wind speeds in remote areas such as satellite remote sensing, water forecasting models and numeric models, identifying limitations and strengths of each set of data. Undoubtedly, with the advent of marine energy-based technologies, good planning based on the factors driving new technologies and the potential of these approaches is an essential step.

First, according to the studies of satellites ability to measure marine renewables data, regional measurements and simulation results with the long-term implementation of numeric models with high spatial resolution, possible effects of wind farm facilities in the oceanography installation site, accurate information on protected areas and marine habitats were understood. Second—according to the identified pros and cons of using these techniques—it supports corporate investment in the development of new indigenous technologies and the harmonization of existing foreign technologies for further use at the installation site. In addition, there will be a support of government for OW farms as a logical solution to support energy self-sufficiency—especially in remote areas of the mother country such as the islands and also a significant step to reduce environmental impacts, promoting OW energy that has evolved in Europe (such as the northern regions) to achieve commercial and financial development. Lastly, since these two dimensions are interconnected, this research underlined that there is a need to clearly understand the framework of equipment and its working principles.

The technology of using satellites to identify suitable areas for setting up wind farms is very promising and requires good planning. This proper planning is possible only with full knowledge of satellite capabilities and different methods for measuring wind sources. For this purpose, it is very useful to have interdisciplinary knowledge and exchange of updated information between different sectors of the development such as universities and companies. For complete understanding, few more factors, such as economic, social, environmental, legal and technological issues for setting up offshore wind farms, must be considered to achieve an appropriate framework. Since all the mentioned steps in the latter stage are interconnected and have a direct impact on each other, so it is important to design a comprehensive path to simultaneously coordinate the study with all effective factors.

Author Contributions: Conceptualization, M.M.N., R.U.S. and D.A.G.; methodology, M.M.N., R.U.S. and D.A.G.; investigation, M.M.N., R.U.S., A.H., A.R., N.A. and D.A.G.; resources, M.M.N., R.U.S., A.H., A.R., N.A. and D.A.G.; data curation, M.M.N., R.U.S., A.H., A.R., N.A. and D.A.G.; writing—original draft preparation, M.M.N., R.U.S. and D.A.G.; supervision, D.A.G. All authors have read and agreed to the published version of the manuscript.

Funding: This research was funded by European Union's Horizon 2020 research and innovation program under grant agreement No 727277 within the project ODYSSEA "Operating a network of integrated observatory systems in the Mediterranean sea".

Acknowledgments: This research was carried out within ODYSSEA project that received funding from the European Union's Horizon 2020 research and innovation program under grant agreement No 727277.

Abbreviations

RS	Remote sensing
SWOT	Strengths, weaknesses, opportunities and threats
RESs	Renewable energy sources
MREs	Marine renewable energies
OW	Offshore wind
WEGs	Wind turbine generators
WECs	Wave energy converters
SAR	Synthetic aperture radar
ECMWF	European center for medium-range weather forecasts
ESA	European space agency
MWS	Mean wind speed
WPD	Wind energy density
SWH	Significant wave height
SCS	Sea current speed
SNAP software	Sentinel application platform
ENVI software	Environment for visualizing images
ROI	Region of interest
GIS	Geographic information system
VV	Vertical vertical
CMOD	C geophysical model function

References

1. Heydari, A.; Astiaso Garcia, D.; Keynia, F.; Bisegna, F.; De Santoli, D. Renewable Energies Generation and Carbon Dioxide Emission Forecasting in Microgrids and National Grids using GRNN-GWO Methodology. *Energy Procedia* **2019**, *159*, 154–159. [CrossRef]
2. Neshat, M.; Nezhad, M.M.; Abbasnejad, E.; Groppi, D.; Heydari, A.; Garcia, D.A.; Tjernberg, L.B.; Alexander, B.; Wagner, M. Hybrid Neuro-Evolutionary Method for Predicting Wind Turbine Power Output. *arXiv* **2020**, arXiv:2004.12794.
3. Lamagna, M.; Nastasi, B.; Groppi, D.; Nezhad, M.M.; Astiaso Garcia, D. Hourly energy profile determination technique from monthly energy bills. *Build. Simul.* **2020**. [CrossRef]
4. Razmjoo, A.; Khalili, N.; Majidi Nezhad, M.; Mokhtari, N.; Davarpanah, A. The main role of energy sustainability indicators on the water management. *Model. Earth Syst. Environ.* **2020**. [CrossRef]
5. Park, E. Social acceptance of green electricity: Evidence from the structural equation modeling method. *J. Clean. Prod.* **2019**, *215*, 796–805. [CrossRef]
6. Ahmadi, A.; Ehyaei, M.A.; Doustgani, A.; El Haj Assad, M.; Hmida, A.; Jamali, D.H.; Kumar, R.; Li, Z.X.; Razmjoo, A. Recent Residential Applications of low-temperature solar collector. *J. Clean. Prod.* **2020**, 123549. [CrossRef]
7. Quero, P.; Chica, J.I.; García, J. Blue energy and marine spatial planning in Southern Europe. *Energy Policy* **2020**, *140*, 111421. [CrossRef]
8. Soukissian, T.H.; Adamopoulos, C.H.; Prospathopoulos, A.; Karathanasi, F.; Stergiopoulou, L. Marine Renewable Energy Clustering in the Mediterranean Sea: The Case of PELAGOS Project. *Front. Energy Res.* **2019**, *7*, 16. [CrossRef]
9. Heydari, A.; Astiaso Garcia, D.; Keynia, F.; Bisegna, F.; De Santoli, L. A novel composite neural network based method for wind and solar power forecasting in microgrids. *Appl. Energy* **2019**, *251*, 113353. [CrossRef]
10. Onea, F.; Ciortan, S.; Rusu, E. Assessment of the potential for developing combined wind-wave projects in the European nearshore. *Energy Explor. Exploit.* **2017**, *28*, 58. [CrossRef]
11. Costoya, X.; Salvador, S.; Carvalho, D.; Sanz-larruga, F.J.; Gimeno, L. An overview of offshore wind energy resources in Europe under present and future climate. *Ann. N. Y. Acad. Sci.* **2019**, *1436*, 70–97. [CrossRef]
12. Hasager, C.B.; Badger, M.; Peña, A.; Larsén, X.G.; Bingöl, F. SAR-Based Wind Resource Statistics in the Baltic Sea. *Remote Sens.* **2011**, *3*, 117–144. [CrossRef]
13. Abanades, J. Wind Energy in the Mediterranean Spanish ARC: The Application of Gravity Based Solutions. *Front. Energy Res.* **2019**. [CrossRef]
14. Soukissian, T.H.; Denaxa, D.; Karathanasi, F.; Prospathopoulos, A.; Sarantakos, K.; Iona, A.; Georgantas, A.; Mavrakos, A. Marine Renewable Energy in the Mediterranean Sea: Status and Perspectives. *Energies* **2017**, *10*, 1512. [CrossRef]
15. Castro-santos, L.; Lamas-galdo, M.I.; Filgueira-vizoso, A. Managing the oceans: Site selection of a floating offshore wind farm based on GIS spatial analysis. *Mar. Policy* **2020**, *113*, 103803. [CrossRef]
16. Kemal, P.; Galparsoro, I.; Depellegrin, D.; Bald, J.; Pérez-morán, G.; Borja, A. A modelling approach for offshore wind farm feasibility with respect to ecosystem-based marine spatial planning. *Sci. Total Environ.* **2019**, *667*, 306–317. [CrossRef]
17. Pimenta, F.M.; Silva, A.R.; Assireu, R.T.; Almeida, V.D.; Saavedra, O.R. Brazil Offshore Wind Resources and Atmospheric Surface Layer Stability. *Energies* **2019**, *12*, 4195. [CrossRef]
18. Windeurope.org. Offshore Wind in Europe, Key Trends and Statistics 2017. 2018. Available online: https://windeurope.org/wp-content/uploads/files/about-wind/statistics/WindEurope-Annual-Offshore-Statistics-2017.pdf (accessed on 14 April 2020).
19. *Offshore Wind Energy: 2020 Mid-Year Statistics*; Windeurope.org; WindEurope: Brussels, Belgium, 2020.
20. Onea, F.; Rusu, E. Efficiency assessments for some state of the art wind turbines in the coastal environments of the Black and the Caspian seas. *Energy Explor. Exploit.* **2016**, *34*, 217. [CrossRef]
21. Peña, A.; Hasager, C.B.; Lange, J.; Anger, J.; Badger, M.; Bingöl, F.; Bischoff, O.; Cariou, J.; Dunne, F.; Emeis, S.; et al. *Remote Sensing for Wind Energy*; (DTU Wind Energy E.; No. 0029(EN)); DTU Wind Energy: Roskilde, Denmark, 2013.
22. Gout, J.P.; Yamaguchi, A.; Ishihara, T. Measurement and Prediction of Wind Fields at an Offshore Site by Scanning Doppler LiDAR and WRF. *Atmosphere* **2020**, *11*, 442. [CrossRef]

23. Chaurasiya, P.K.; Ahmed, S.; Warudkar, V. Comparative analysis of Weibull parameters for wind data measured from met-mast and remote sensing techniques. *Renew. Energy* **2018**, *115*, 1153–1165. [CrossRef]

24. Byrne, B.; Houlsby, G.; Martin, C.; Fish, P. Suction Caisson Foundations for Offshore Wind Turbines. *Wind Eng.* **2020**, *26*, 145–155. [CrossRef]

25. Khan, K.S.; Tariq, M. Wind resource assessment using SODAR and meteorological mast—A case study of Pakistan. *Renew. Sustain. Energy Rev.* **2018**, *81*, 2443–2449. [CrossRef]

26. Goffetti, G.; Montini, M.; Volpe, F.; Gigliotti, M.; Pulselli, F.; Sannino, G.; Marchettini, N. Disaggregating the SWOT Analysis of Marine Renewable Energies. *Front. Energy Res.* **2018**, *6*, 1–11. [CrossRef]

27. Pisacane, G.; Sannino, G.; Carillo, A.; Struglia, M.V.; Bastianoni, S. Marine Energy Exploitation in the Mediterranean Region: Steps Forward and Challenges. *Front. Energy Res.* **2018**. [CrossRef]

28. Nikolaidis, G.; Karaolia, A.; Matsikaris, A.; Nikolaidis, A. Blue Energy Potential Analysis in the Mediterranean. *Front. Energy Res.* **2019**, *7*, 1–12. [CrossRef]

29. Azzellino, A.; Lanfredi, C.; Riefolo, L.; De Santis, V. Combined Exploitation of Offshore Wind and Wave Energy in the Italian Seas: A Spatial Planning Approach. *Front. Energy Res.* **2019**, *7*, 42. [CrossRef]

30. Kerbaol, V.; Chapron, B.; Vachon, P.W. Analysis of ERS-1/2 synthetic aperture radar wave mode imagettes. *J. Geophys. Res. Ocean.* **1998**, *103*, 7833–7846. [CrossRef]

31. Ren, Y.; Li, X.M.; Zhou, G. Sea surface wind retrievals from SIR-C/X-SAR data: A revisit. *Remote Sens.* **2015**, *7*, 3548–3564. [CrossRef]

32. Sentinel-1 Product Specification. Ref: S1-RS-MDA-52-7441. Available online: https://sentinel.esa.int/documents/247904/1877131/Sentinel-1-Product-Specification (accessed on 13 August 2020).

33. Karagali, I.; Badger, M.; Hahmann, A.N.; Peña, A.; Hasager, C.B.; Maria, A. Spatial and temporal variability of winds in the Northern European Seas. *Renew. Energy* **2013**, *57*, 200–210. [CrossRef]

34. Bentamy, A.; Croize-fillon, D. Spatial and temporal characteristics of wind and wind power off the coasts of Brittany. *Renew. Energy* **2014**, *66*, 670–679. [CrossRef]

35. Signell, R.P.; Chiggiato, J.; Horstmann, J.; Doyle, J.D.; Pullen, J.; Askari, F. High-resolution mapping of Bora winds in the northern Adriatic Sea using synthetic aperture radar. *J. Geophys. Res.* **2010**, *115*, C04020. [CrossRef]

36. Agust, A.; Vasiljevic, D.; Beljaars, A.; Bock, O.; Guichard, F.; Nuret, M.; Mendez, A.G.; Andersson, E.; Bechtold, P.; Fink, A.; et al. Radiosonde humidity bias correction over the West African region for the special AMMA reanalysis at ECMWF. *Q. J. R. Meteorol. Soc.* **2009**, *617*, 595–617. [CrossRef]

37. Hersbach, H. Comparison of C-Band Scatterometer CMOD5. N Equivalent Neutral Winds with ECMWF. *J. Atmos. Ocean. Technol.* **2010**, *27*, 721–736. [CrossRef]

38. Anderson, P.S.; Ladkin, R.S.L.; Renfrew, I.A. An autonomous Doppler sodar wind profiling system. *J. Atmos. Ocean. Technol.* **2005**, *22*, 1309–1325. [CrossRef]

39. Frank, C.W.; Pospichal, B.; Wahl, S.; Keller, J.D.; Hense, A.; Crewell, S. The added value of high resolution regional reanalyses for wind power applications. *Renew. Energy* **2019**, *148*, 1094–1109. [CrossRef]

40. Arun, S.V.V.; Nagababu, G.; Kumar, R. Comparative study of offshore winds and wind energy production derived from multiple scatterometers and met buoys. *Energy* **2019**, *185*, 599–611. [CrossRef]

41. Curto, D.; Franzitta, V.; Trapanese, M.; Cirrincione, M. A Preliminary Energy Assessment to Improve the Energy Sustainability in the Small Islands of the Mediterranean Sea. *J. Sustain. Dev. Energy Water Environ. Syst.* **2005**. [CrossRef]

42. Meschede, H.; Holzapfel, P.; Kadelbach, F.; Hesselbach, J. Classification of global island regarding the opportunity of using RES. *Appl. Energy* **2016**, *175*, 251–258. [CrossRef]

43. Rusu, E.; Onea, F. An assessment of the wind and wave power potential in the island environment. *Energy* **2019**, *175*, 830–846. [CrossRef]

44. Onea, F.; Rusu, E. The Expected Shoreline Effect of a Marine Energy Farm Operating Close to The Expected Shoreline Effect of a Marine Energy Farm Operating Close to Sardinia Island. *Water* **2019**, *11*, 2303. [CrossRef]

45. Vicinanza, D.; Cappietti, L.; Ferrante, V.; Contestabile, P. Estimation of the wave energy in the Italian offshore. *J. Coast. Res.* **2011**, *64*, 613–617.

46. Davy, R.; Gnatiuk, N.; Pettersson, L.; Bobylev, L. Climate change impacts on wind energy potential in the European domain with a focus on the Black Sea. *Renew. Sustain. Energy Rev.* **2018**, *81*, 1652–1659. [CrossRef]

47. Ulazia, A.; Saenz, J.; Ibarra-berastegui, G. Sensitivity to the use of 3DVAR data assimilation in a mesoscale model for estimating offshore wind energy potential. A case study of the Iberian northern coastline. *Appl. Energy* **2016**, *180*, 617–627. [CrossRef]

48. Remmers, T.; Cawkwell, F.; Desmond, C.J.; Murphy, J. The potential of advanced scatterometer (ASCAT) 12.5 KM costal observations for offshore wind farm site selection in irish waters. *Energies* **2019**, *12*, 206. [CrossRef]

49. Westerberg, V.; Jacobsen, J.B.; Lifran, R. Offshore wind farms in Southern Europe—Determining tourist preference and social acceptance. *Energy Res. Soc. Sci.* **2015**, *10*, 165–179. [CrossRef]

50. Brudermann, T.; Zaman, R.; Posch, A. Not in my hiking trail? Acceptance of wind farms in the Austrian Alps. *Clean Technol. Environ. Policy* **2019**, *21*, 1603–1616. [CrossRef]

51. Sustainable Tourism in the Mediterranean: State of Play and Strategic Directions. May 2017. Available online: https://planbleu.org/sites/default/files/publications/cahier17_tourisme_en_web.pdf (accessed on 12 August 2020).

52. Scolozzi, R.; Schirpke, U.; Morri, E.; Amato, D.D.; Santolini, R. Ecosystem services-based SWOT analysis of protected areas for conservation strategies. *J. Environ. Manag.* **2020**, *146*, 543–551. [CrossRef]

53. Saidur, R.; Rahim, N.A.; Islam, M.R.; Solangi, K.S. Environmental impact of wind energy. *Renew. Sustain. Energy Rev.* **2011**, *15*, 2423–2430. [CrossRef]

54. Goodale, M.W.; Milman, A. Assessing the cumulative exposure of wildlife to off shore wind energy development. *J. Environ. Manag.* **2019**, *235*, 77–83. [CrossRef]

55. Jiang, D.; Zhuang, D.; Huang, Y.; Wang, J.; Fu, J. Evaluating the spatio-temporal variation of China's offshore wind resources based on remotely sensed wind field data. *Renew. Sustain. Energy Rev.* **2013**, *24*, 142–148. [CrossRef]

56. Guo, Q.; Huang, R.; Zhuang, L.; Zhang, K. Assessment of China' s Offshore Wind Resources Based on the Integration of Multiple Satellite Data and Meteorological Data. *Remote Sens.* **2019**, *11*, 2680. [CrossRef]

57. Nezhad, M.M.; Groppi, D.; Marzialetti, P.; Fusilli, L.; Laneve, G.; Cumo, F. Wind energy potential analysis using Sentinel-1 satellite: A review and a case study on Mediterranean islands. *Renew. Sustain. Energy Rev.* **2019**, *109*, 499–513. [CrossRef]

58. Veci, L. SNAP Command Line Tutorial Graph Processing. ESA 2016. 2016. Available online: https://step.esa.in (accessed on 14 August 2020).

59. ENVI Tutorial: Classification Methods. Available online: http://www.harrisgeospatial.com (accessed on 13 August 2020).

60. ENVI Tutorial: Introduction to ENVI Table of Contents. Available online: http://webcache.googleusercontent.com/search?q=cache:qsOJnZFMiicJ:faculty.wwu.edu/wallin/envr442/ENVI/ENVI_Intro.pdf+&cd=2&hl=en&ct=clnk&gl=us (accessed on 14 August 2020).

61. Díaz, P.; Markus, C.; Javier, B.; Bravo, D.; Schardinger, I. Developing a wind energy potential map on a regional scale using GIS and multi—Criteria decision methods: The case of Cadiz (South of Spain). *Clean Technol. Environ. Policy* **2018**, *20*, 1167–1183. [CrossRef]

62. Díaz, P.; Javier, C.; Bravo, D.; Prieto, A. Integrating MCDM and GIS for renewable energy spatial models: Assessing the individual and combined potential for wind, solar and biomass energy in Southern Spain. *Clean Technol. Environ. Policy* **2019**, *21*, 1855–1869. [CrossRef]

63. Rachid, G.; El Fadel, M. Comparative SWOT analysis of strategic environmental assessment systems in the Middle East and North Africa region. *J. Environ. Manag.* **2013**, *125*, 85–93. [CrossRef]

64. Piasecki, A.; Jurasz, J.; Kies, A. Measurements and reanalysis data on wind speed and solar irradiation from energy generation perspectives at several locations in Poland. *SN Appl. Sci.* **2019**, *1*, 865. [CrossRef]

65. Michele, F.; Adamo, M.; Lucas, R.; Blonda, P. Sea surface wind retrieval in coastal areas by means of Sentinel-1 and numerical weather prediction model data. *Remote Sens. Environ.* **2018**, *225*, 379–391.

66. Badger, M.; Ahsbahs, T.; Maule, P.; Karagali, I. Remote Sensing of Environment Inter-calibration of SAR data series for o ff shore wind resource assessment. *Remote Sens. Environ.* **2019**, *232*, 111316. [CrossRef]

67. Ahsbahs, T.; Maclaurin, G.; Draxl, D.; Jackson, C.; Monaldo, F. US East Coast synthetic aperture radar wind atlas for offshore wind energy. *Wind Energy Sci.* **2019**. [CrossRef]

68. Nezhad, M.M.; Heydari, A.; Groppi, D.; Cumo, F.; Garcia, D.A. Wind source potential assessment using Sentinel 1 satellite and a new forecasting model based on machine learning: A case study Sardinia islands. *Renew. Energy* **2020**, *155*, 212–224. [CrossRef]

69. Wang, Y.H.; Walter, R.K.; White, C.; Farr, H.; Ruttenberg, B.I. Assessment of surface wind datasets for estimating offshore wind energy along the Central California Coast. *Renew. Energy* **2018**, *133*, 343–353. [CrossRef]

70. Guo, Q.; Xu, X.; Zhang, K.; Li, Z.; Huang, W.; Mansaray, L.R.; Liu, W.; Wang, X.; Gao, J.; Huang, J. Assessing Global Ocean Wind Energy Resources Using Multiple Satellite Data. *Remote. Sens.* **2018**, *10*, 100. [CrossRef]

71. Barthelmie, R.J.; Pryor, S.C. Can Satellite Sampling of Offshore Wind Speeds Realistically Represent Wind Speed Distributions ? *J. Appl. Meteorol.* **2003**, *42*, 83–94. [CrossRef]

72. Dee, D.P.; Uppala, S.M.; Simmons, A.J.; Berrisford, P.; Poli, P.; Kobayashi, S.; Andrae, U.; Balmaseda, M.A.; Balsamo, G.; Bauer, P.; et al. The ERA-Interim reanalysis: Configuration and performance of the data assimilation system. *R. Meteorol. Soc.* **2011**, *137*, 553–597. [CrossRef]

73. Ganea, D.; Mereuta, E.; Rusu, L. Estimation of the Near Future Wind Power Potential. *Energies* **2018**, *11*, 3198. [CrossRef]

74. Fokaides, P.A.; Miltiadous, I.; Neophytou, M.K.; Spyridou, L.P. Promotion of wind energy in isolated energy systems: The case of the Orites wind farm. *Clean Technol. Environ Policy* **2014**, *16*, 477–488. [CrossRef]

75. Wei, X.; Duan, Y.; Liu, Y.; Jin, S.; Sun, C. Onshore-offshore wind energy resource evaluation based on synergetic use of multiple satellite data and meteorological stations in Jiangsu Province, China. *Front. Earth Sci.* **2019**, *13*, 132–150. [CrossRef]

76. Elsner, P. Continental-scale assessment of the African offshore wind energy potential: Spatial analysis of an under-appreciated renewable energy resource. *Renew. Sustain. Energy Rev.* **2019**, *104*, 394–407. [CrossRef]

77. Pantusa, D.; Tomasicchio, G.R. Large-scale offshore wind production in the Mediterranean Sea. *Cogent Eng.* **2019**, *6*, 1661112. [CrossRef]

78. Bruun Christiansen, M.; Koch, W.; Horstmann, J.; Bay Hasager, C.; Nielsen, M. Wind resource assessment from C-band SAR. *Remote Sens. Environ.* **2006**, *105*, 68–81. [CrossRef]

79. Rose, S.; Apt, J. What can reanalysis data tell us about wind power? *Renew. Energy* **2015**, *83*, 963–969. [CrossRef]

80. Olauson, J. ERA5: The new champion of wind power modelling ? *Renew. Energy* **2018**. [CrossRef]

81. Arun, S.V.V.; Nagababu, G.; Sharma, R.; Kumar, R. Synergetic use of multiple scatterometers for offshore wind energy potential assessment. *Ocean Eng.* **2020**, *196*, 106745. [CrossRef]

82. Nezhad, M.M.; Groppi, D.; Marzialetti, P.; Laneve, G. A sediment detection analysis with multi sensor satellites: Caspian sea and persian gulf case studies. In Proceedings of the 4th World Congress on Civil, Structural, and Environmental Engineering, CSEE, Rome, Italy, 7–9 April 2019. [CrossRef]

83. Korsbakken, E.; Johannessen, J.A.; Johannessen, O.M. Coastal wind field retrievals from ERS synthetic aperture radar images. *J. Geophys. Res.* **1998**. [CrossRef]

84. Alpers, W. Katabatic wind fields in coastal areas studied by ERS-1 synthetic aperture radar imagery and numerical modeling coastal. *J. Geophys. Res.* **1998**, *103*, 7875–7886. [CrossRef]

85. Nezhad, M.M.; Groppi, D.; Marzialetti, P.; Piras, G.; Laneve, G. Mapping Sea Water Surface in Persian Gulf, Oil Spill Detection Using Sentinal-1 Images. In Proceedings of the 4th World Congress on New Technologies, Madrid, Spain, 19–21 August 2018. [CrossRef]

86. Ribal, A.; Young, I.R. 33 years of globally calibrated wave height and wind speed data based on altimeter observations. *Sci. Data* **2019**, 1–15. [CrossRef]

87. Hasager, C.B.; Badger, M.; Nawri, N.; Furevik, B.R. Mapping Offshore Winds Around Iceland Using Satellite Synthetic Aperture Radar and Mesoscale Model Simulations. *IEEE J. Sel. Top. Appl. Earth Obs. Remote Sens.* **2015**, *8*, 5541–5552. [CrossRef]

88. Hasager, C.B. Offshore winds mapped from satellite remote sensing. *Wiley Interdiscip. Rev. Energy Environ.* **2014**, *3*, 594–603. [CrossRef]

89. Smyth, K.; Christie, N.; Burdon, D.; Atkins, J.P.; Barnes, R.; Elliott, M. Renewables-to-reefs?—Decommissioning options for the offshore wind power industry. *Mar. Pollut. Bull.* **2015**, *90*, 247–258. [CrossRef]

90. Bergström, L.; Kautsky, L.; Malm, T.; Rosenberg, R. Effects of offshore wind farms on marine wildlife—A generalized impact assessment. *Environ. Res. Lett.* **2014**, *9*, P034012. [CrossRef]

Appl. Sci. **2020**, *10*, 6398

91. Zhao, X.; Ren, L. Focus on the development of offshore wind power in China: Has the golden period come? *Renew. Energy* **2020**, *81*, 644–657. [CrossRef]
92. Carvalho, D.; Rocha, A.; Gómez-Gesteira, M.; Santos, C.S. Offshore winds and wind energy production estimates derived from ASCAT, OSCAT, numerical weather prediction models and buoys—A comparative study for the Iberian Peninsula Atlantic coast. *Renew. Energy* **2017**, *102*, 433–444. [CrossRef]

Article

Dynamic Evaluation of Heat Thefts Due to Different Thermal Performances and Operations between Adjacent Dwellings

Laura Canale [1,2], Vittoria Battaglia [1], Giorgio Ficco [2], Giovanni Puglisi [3] and Marco Dell'Isola [2,*

[1]   Dipartimento di Ingegneria, Università degli Studi di Napoli "Parthenope", 80143 Napoli, Italy; l.canale@unicas.it (L.C.); vittoria.battaglia@uniparthenope.it (V.B.)

[2]   Dipartimento di Ingegneria Civile e Meccanica, Università degli Studi di Cassino e del Lazio Meridionale, Via G. Marconi 10, 03043 Cassino, Italy; ficco@unicas.it

[3]   ENEA Centro della Casaccia, Via Anguillarese 301, 00123 Rome, Italy; giovanni.puglisi@enea.it

*   Correspondence: dellisola@unicas.it

Received: 2 March 2020; Accepted: 30 March 2020; Published: 2 April 2020

Abstract: Apartment position and operation within buildings play a significant role on energy consumption and also on perceived thermal comfort. Dwellings with favorable positions can have significant benefit, also when heated for a limited number of hours, if compared to apartments located in disadvantaged positions (i.e., upper or lower floors or north-oriented). This may be the cause of debates, especially in buildings with central heating, when heat costs are shared among tenants by means of sub-metering systems. In this paper, authors address this issue by studying the "heat thefts" phenomenon in dynamic conditions in a low-insulated building, when the heating system is used unevenly by the tenants (i.e., with different temperatures and/or use). To this end, a social housing building located in Mediterranean climate, where daily temperature excursions and solar heat gains enhance the dynamics of the heat flows, has been chosen as the case-study. The real operation of the building has been simulated in different operational scenarios and the model has been validated against energy consumption data collected experimentally. Results confirm that special allocation and or/compensation strategies should be taken in heat costs allocation in order to avoid accentuating situations of inequalities, especially in low-insulated and/or occasionally heated buildings.

Keywords: stolen heat; energy efficiency; heat metering; heat accounting; dynamic simulation; social housing; building simulation; heat cost allocators; thermostatic radiator valves

---

1. Introduction

As well known, in recent years targeting energy efficiency in residential buildings has become a key issue for the European Union (EU) which has committed itself to cut its greenhouse gas emissions to 80–95% below 1990 levels by 2050 [1]. Indeed, buildings are responsible for 36% of global final energy consumption in Europe and for nearly 40% of total direct and indirect CO_2 emissions [2], of which about 70% is due to heating the existing building stock. For this reason, in 2012 the EU issued the so-called "Energy Efficiency Directive" (EED) [3] amended in 2018 as part of the "Clean energy for all Europeans package", establishing a set of binding measures to help the EU reach its energy efficiency target. In particular, Article 9 highlights the need to ensure that final customers are provided with meters that accurately reflect their actual energy consumption and set as mandatory the installation of individual meters in buildings supplied from a central source, whether sub-metering is cost-efficient or the related costs are proportionate in relation to the potential energy savings [4]. In this way, end-users should be aware, and at the same time responsible, of their energy consumption.

This obligation has led to a number of issues related to a fair billing service [5,6]. Indeed, fair and transparent heat cost allocation for heating/cooling/hot water costs is a difficult subject.

In fact, especially in old and existing buildings, the lack of insulation of walls can generate heat transfers between adjacent apartments, when these are heated at different set-point temperatures (the so-called "stolen heat" or "heat thefts"), being the cause of involuntary over-consumptions for dwelling surrounded by numerous unheated or occasionally heated apartments. In this sense, a number of authors addressed the problem, with some [7–9] from the perspective of heat cost allocation, others [10–13] analyzing the phenomenon in a quantitative manner.

Within the first group, Liu et al. [7] presented an alternative method for allocating heat costs in multi-apartment buildings based on measuring the on-off ratio of the thermostatic valves on the radiators. A cost-efficient method was developed for reallocation of heating costs based on heat transfers between the adjacent apartments by Siggelsten [14] which was then further developed by Michnikowski [9] basing on the data of the average indoor temperature provided by special heat cost allocators.

Among the second group, Gafsi and Lefebvre [10] demonstrated that an apartment can gain up to 90% of heat from adjacent apartments and highlighted the complexity of the phenomenon and the large number of influence factors. Pakanen and Karjalainen [11] proposed a method for estimating static heat flows between adjacent rooms in a hypothetical environment simulated using TRNSYS software. Lukić et al. [12] showed that, in fully insulated buildings, an unheated dwelling can steal about 80% of its total energy need from the surrounding apartments, by simulating a real building using EnergyPlus simulation software. In [13], an energy simulation is adopted to calculate the heat transfer proportion with the validation of on-site measurement, showing an unexpected amount of adjacent room heat transfer up 70% of the total heating load.

That said, it is clear that apartment location and operation can play a significant role on its energy consumption. In fact, depending on the type of building, dwellings with favorable positions, such as those with a high interior to total wall surface ratio, the ones south oriented, etc., can have significant benefit by gaining heat from surrounding dwellings, if compared to apartments located in more disadvantaged positions.

In this paper, the issue of heat thefts has been investigated by performing a dynamic simulation on a real case-study represented by a social housing building supplied by a centralized natural gas boiler, using TRNSYS simulation software. To this end, the real operation of the building (base scenario) has been simulated and the model has been validated against real energy consumption data collected by means of direct and indirect heat metering systems installed. Thanks to the validated model, it was possible to simulate two additional scenarios (scenarios a and b) in which two apartments representative of the most and the least favored positions were considered, individually, unheated.

The main aim of this work is to analyze different scenarios of building operation, with particular reference to the effects of heat transfers between adjacent dwellings in terms of energy consumption and heat cost allocation under uneven use of the heating system (in terms of occupancy of the dwellings and different set-point temperatures). The dynamic of heat transfers has been simulated with reference to an Italian building with low thermal energy performances and in a Mediterranean climate, to highlight the differences with other existing studies. Thus, the novelty of this work relies on the fact that the authors focused on the issue of heat thefts in buildings from the point of view of the allocation of heating costs among the dwelling units. This allowed for greater understanding of the dynamic behavior of the heat fluxes in relation to the single apartments rather than the entire building, especially considering that the Mediterranean climatic conditions may enhance the dynamics of the heat flows due to temperature excursions and solar heat gains during the day.

The main findings of the analysis revealed dynamic effects of "inversion" of the heat transfers between adjacent apartments, that would not have been evident in static conditions, and that should be carefully taken into account for heat cost allocation in social housing context. The results may be particularly useful because the case-study is representative of the Italian building stock in terms of constructive characteristics and heating plant and also because the influence of solar heat gains in Mediterranean climate conditions on the heat theft phenomenon is highlighted through the dynamic simulation.

2. Materials and Methods

2.1. The Building Case-Study

The object of this study is represented by a low-insulated building located in a Mediterranean climate. To this end, a social housing building located in the province of Frosinone, Italy has been chosen as the subject of this study and simulated under both the hypothesis of uneven operational temperatures and occasional heating. In particular, the building has been studied under actual operational conditions, meaning that real set-point temperatures and operation hours of the heating system effectively set by the users were employed for the simulation. Specifically, the set-point temperatures of the fully-occupied building simulation varied between 18.6 °C and 22.1 °C as a function of the given end-user, while the operation hours were equal for all the dwellings (i.e., from 6 to 8 a.m. and from 15 to 22 p.m. according to the real settings of the centralized heating plant). The outdoor climatic conditions are those typical of the climatic zone in which the building is located and vary between (23.2/−2.1 °C) in the simulation period.

The building was built in 1979 by ATER, the Italian Territorial Agency for Social Housing Buildings. The hydronic heating system consists on a central natural gas boiler with a maximum power of 152 kW, located in the ground floor, with uninsulated distribution pipes running mainly in the outer walls in vertical configuration [15] and emission terminals consisting of traditional cast iron radiators.

Nine dwellings are arranged in two blocks each of which with three floors. The first, consisting of three dwellings, is located above the front porch (type C); the second, located above the garages, is made up of six dwellings (two on each floor), three north-west oriented (type B) and three south-west oriented (type A). Figure 1a,b shows, respectively, the cross section of the building and the plant scheme of a representative floor, while Figure 2 shows a picture of the investigated building.

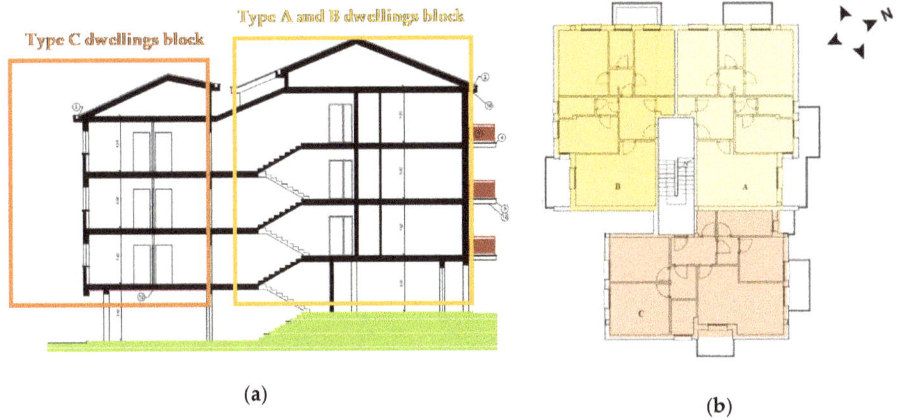

(a)

(b)

Figure 1. (a) Cross-section of the building case-study; (b) plant scheme of a representative floor.

Dwellings of types A and B have a net floor area of about 79 m^2, while type C ones have a net floor area of about 86 m^2. The net interstorey height is about 2.7 m.

It is underlined that almost all the walking surface of the first floor, which is located above the garages and the porch, is exposed to the external environment (the garages have open doors), as well as almost all the ceiling of the last floor faces the unheated attic.

Figure 2. Picture of the building case-study.

The main thermal and physical characteristics of the investigated building, collected through on-site inspections and document review, are listed in Table 1. A distinction has been made between two different types of floors, since the outer floor of the dwelling located on the porch has undergone a thermal insulation with the addition of an external layer of expanded polystyrene 0.05 m thick, while the outer floor of the other apartments located at the first floor did not.

Table 1. Main constructive characteristics of the investigated building.

Building Element	Layers (from Inside to Outside)	Thickness [m]	Thermal Transmittance [W/(m^2 K)]
Outer ceiling	Lime/gypsum plaster	0.020	2.68
	Concrete	0.200	
	Waterproofing layer	0.004	
	Tiles	0.015	
Outer floor (porch apt.)	Floor tiles	0.010	0.49
	Lean concrete	0.050	
	Hollow core concrete	0.180	
	Expanded polystyrene	0.050	
	Lime/gypsum plaster	0.020	
Outer vertical walls	Lime/gypsum plaster	0.020	1.25
	Hollow clay bricks	0.100	
	Air Gap	0.080	
	Hollow concrete bricks	0.100	
Inner vertical walls	Lime/gypsum plaster	0.010	2.10
	Hollow clay bricks	0.100	
	Lime/gypsum plaster	0.010	
Inner ceiling and all other floors	Floor tiles	0.010	1.25
	Lean concrete	0.050	
	Hollow core concrete	0.180	
	Lime/gypsum plaster	0.020	

In September 2017, two-sensors—Heat Cost Allocators (HCAs) and Thermostatic Radiator Valves (TRVs)—were installed on each radiator to comply with the EED obligation [16], transposed in Italy with the Legislative Decree 102/2014 and subsequent amendments and integrations [17]. To allow an optimal system operation, a variable speed circulation pump was installed in the central boiler room. Together with the HCAs and the TRVs, two temperature sensors were installed in the bedroom and in the living room of each dwelling, which are able to measure the daily average temperature of the room as an average of 24-h measurements with a 6-min acquisition frequency. Additionally, a main thermal energy meter was installed on the main distribution pipes downstream of the centralized boiler in order to measure the total energy consumption of the building.

The installed heating powers have been estimated through the dimensional method [18,19], after mapping the dimensions of all radiators, which was also required as preliminary action for the purposes of programming the HCAs. In the context of a specific project, survey questionnaires have been administrated to the inhabitants of the investigated building and on-field informative meetings have been organized, aimed at designing a suitable feedback strategy to enhance end-user's awareness. In particular, the analysis of surveys allowed authors to collect specific information about behavioral features which have been also meaningful for the present analysis, such as the ability to use thermostatic valves and chrono thermostat, and their feeling about indoor temperature in the apartment [20].

Table 2 contains the data related to the calculated installed heating power and the number of inhabitants of each apartment of the building case study.

Table 2. Installed heating power and number of inhabitants of each dwelling.

Dwelling Number	Installed Power [kW]	Number of Inhabitants
1C	12.96	1
2A	10.17	2
3B	10.63	3
4C	9.78	2
5A	9.05	2
6B	8.67	7
7C	13.80	4
8A	11.53	1
9B	11.99	2

2.2. Simulation Model

A dynamic simulation of the case-study building has been performed by the authors by means of the energy tool TRNSYS-17 [21], which allows to simulate the dynamic behavior multi-zone buildings and their systems. The software carries out the simulation basing on a thermal balance (typically on an hourly basis) taking into account the effects of the accumulation and of the thermal release of the opaque building envelope [22,23].

Both the building envelope and the heating system were modeled and simulated and the results were validated according to the real thermal energy measured at both the building level (through the main thermal energy meter) and at the dwelling level (by means of the data collected by the HCAs installed in the dwellings). The heating season 2017–2018 (approximately from November 1st to April 15th) has been chosen as reference period for the validation.

The 3-D building-type model was developed through the Google SketchUp plug-in "Trnsys3d" [24], as shown in Figure 3, in which the thermo-physical properties associated to the building envelope are the ones collected in Table 1.

Due to technical reasons (in Trnsys3d thermal zones must be convex), 24 thermal zones have been used to model the apartments, namely: 3 for type B and type C dwellings and 2 for type A dwellings. For the attic, 8 thermal zones with pitched roof have been implemented. The stairwell has been represented with one thermal zone for each floor.

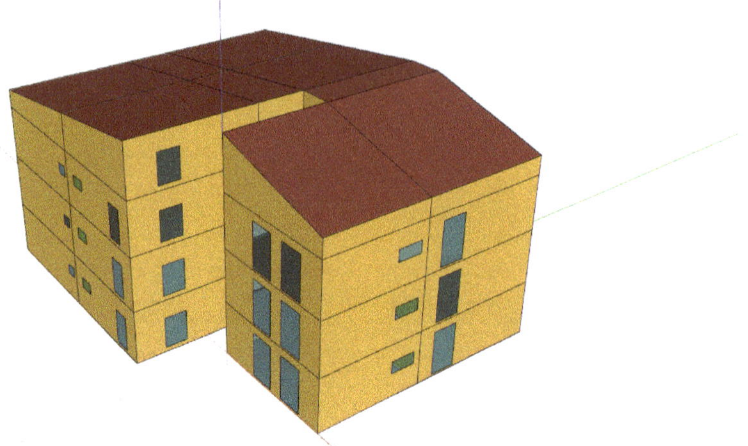

Figure 3. 3-D model of the building case-study.

The infiltration rate (i.e., the number of air volumes in the unit of time entering the building from the outside through the windows due to their inadequate closure) has been set according to the conventional values for residential buildings [25], which depend on air tightness but also on wind speed.

The only considered internal heat gain contribution was the one coming from the inhabitants, since most of the apartments are not equipped with highly energy-consuming electrical devices and only a few of the rooms are actually used by the inhabitants.

The heating system has been modeled as the summation of four components by means of specific system-types, accounting for: (*i*) the emission system, whose heating power represents an input and is given in Table 2; (*ii*) the generation system, whose efficiency is an input parameter and has been calculated as the average generation efficiency of two consecutive years measured during the experimental campaign (i.e., the ratio between the thermal energy consumption measured by the direct heat meter installed downstream to the boiler and the natural gas consumption expressed in kWh and measured by the building natural gas meter); (*iii*) the system circulators; (*iv*) distribution and mixing systems.

Given the presence of TRVs on the emission terminals, with which the inhabitants can adjust the indoor temperature of the rooms, a fluctuant set-point temperature has been associated to each thermal zone. This is represented by the average weekly value assumed by the mean temperature measured through the T-logger sensors installed in the dwellings. The operation of the generation system is controlled by a daily and a seasonal control schedule, simulated, respectively, according to the actual on-off schedule of the heating system (from 6 to 8 a.m. and from 15 to 22 p.m.) and to the heating period established according to the Italian law [26] for the reference climatic zone (climatic zone: D, heating degree-days (HDD): 1911, heating period: 1st November–15th April).

3. Results and Discussion

For each thermal zone, the heating demand was obtained for the base scenario, i.e., the simulation of the full operation of the building, where all apartments are heated with a fluctuating set-point temperature determined as the mean weekly temperature measured by the temperature sensors installed in the dwellings.

Figures 4–6 show the hourly distribution of the heating load for the analyzed dwellings obtained by the dynamic simulation, each representing a macro thermal-zone resulting from the summation of the thermal zones simulating the dwellings. For simplicity reasons, the graphs show only the heating season (i.e., from 1st January to 15th April, and from 1st November to 31st December).

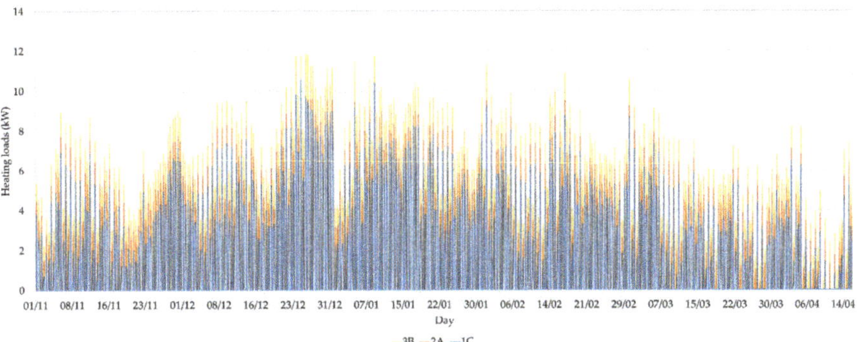

Figure 4. Simulated heating load of apartments 1C, 2A and 3B.

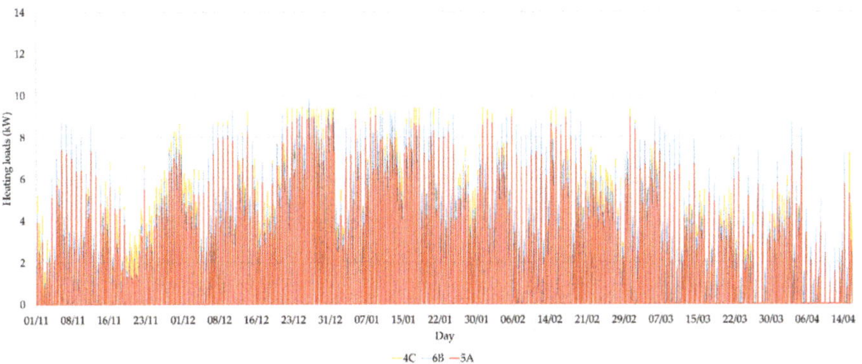

Figure 5. Simulated heating load of apartments 4C, 5A and 6B.

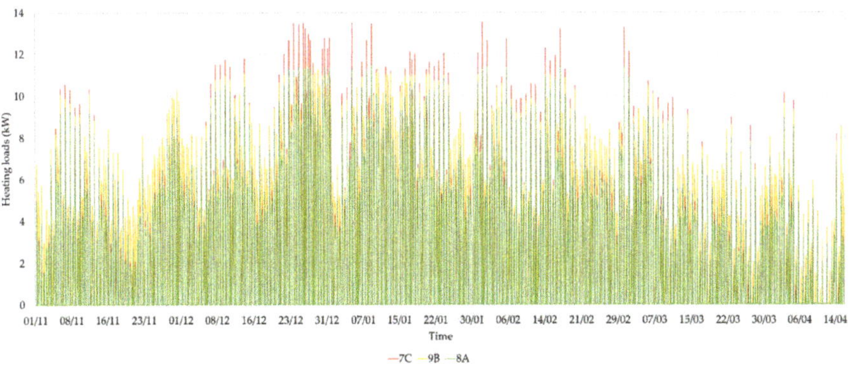

Figure 6. Simulated heating load of apartments 7C, 8A and 9B.

The annual heating consumption was calculated as the area below the power curve obtained by the simulation. Both the estimated and the measured energy consumptions were normalized with respect to the reference HDDs of the building location (1911) as reported by Equations (1) and (2).

$$(normalized\ consumption)_{estimated} = \frac{estimated\ consumption}{HDD_{(weather\ file)}} \cdot 1911 \tag{1}$$

$$(normalized\ consumption)_{measured} = \frac{measured\ consumption}{HDD_{2017-18}} \cdot 1911 \tag{2}$$

where $HDD_{(weather\ file)}$ are the HDDs calculated for the weather file used for the simulation and $HDD_{2017-18}$ are the HDDs of the heating season of 2017–2018 of the building location obtained by the local weather database.

A calibration of the model has been performed, aimed at decreasing the magnitude of the simulation error in some apartments. In fact, after the first simulation, apartments 5A, 6B, 7C and 8A showed the highest simulation errors (see Table 3). In this regard, information was retrieved by the authors in the field, during meetings specifically organized to collect data about end-user's behavior. In particular, it has been found that users of those apartments used to open their windows during the heating hours, instead of employing the TRVs, to decrease the indoor temperature. For this reason, for the above-mentioned apartments, a calibration has been performed by the authors by increasing the ventilation rate until the error decreased to a range of ±15%.

Table 3. Results of validation before and after the calibration process (respectively BC and AC).

Dwelling Number	Estimated Normalized Energy Consumption, BC [kWh]	Estimated Normalized Energy Consumption, AC [kWh]	Measured Normalized Energy Consumption [kWh]	Error [%]	
				BC	AC
1C	4918	5424	5673	−13%	−4%
2A	6578	6847	6796	−3%	1%
3B	7686	7704	7334	5%	5%
4C	6600	7351	7685	−14%	−4%
5A	4552	5934	5578	−18%	6%
6B	4470	7151	7386	−39%	−3%
7C	5846	9015	9617	−39%	−6%
8A	5206	7885	8919	−42%	−12%
9B	8042	9882	9460	−15%	4%
TOTAL	53897	67193	68449	−21%	−2%

Results are reported in Table 3, namely: the normalized measured energy consumption of each dwelling for the heating season 2017–2018 (estimated basing on the reading of the HCAs); the normalized estimated energy consumption (i.e., obtained by means of the dynamic simulation implemented using TRNSYS17), respectively before and after the calibration process (BC and AC); the error between the estimated and the measured normalized thermal energy consumption before and after the calibration process (BC and AC), calculated as per Equation (3).

$$error\ [\%] = \frac{estimated\ consumption - measured\ consumption}{measured\ consumption} \cdot 100 \tag{3}$$

In Table 3, only the energy consumptions normalized with respect to the reference HDDs of the building location are reported.

As shown in Table 3, following the calibration process on dwellings 5A, 6B, 7C and 8A, the error calculated with respect to the total energy consumption of the building decreased to −2%, while the one of single dwellings decreased to a range of +6/−12%. Thus, the validation results have been considered to be acceptable for the purpose of the present analysis.

Indeed, these small deviations may be caused by different causes, as following:

i   potential differences between real and estimated thermal transmittances and thermal mass of the building;

ii   for the purposes of the present analysis, the set-point temperature is set as average measured weekly temperature of the entire thermal zone;

iii   the energy consumption of single dwellings is estimated based on HCAs readings, which depend also on the estimation of radiators' thermal output performed through the dimensional method described in [18,19]. It has been demonstrated that different installation conditions can lead to deviations between operating and standard radiators' thermal output in a range of about 5–15% [27].

The heat flowing through the adjacent wall (Q_h [Wh]) of two dwellings in one hour, h, has been calculated as per Equation (4):

$$Q_h = U{\cdot}A{\cdot}(T_{in} - T_{out}){\cdot}h \tag{4}$$

where U, [W m^{-2} K^{-1}] is the thermal transmittance of the wall, A [m^2] is the area of the wall and $T_{in} - T_{out}$, [K] is the temperature difference between the inner and the outer layer of the wall (with T_{in} indicating the temperature measured at the inner surface—i.e., the surface of the analyzed dwelling—while T_{out} the outer surface temperature—i.e., the temperature of the adjacent dwelling).

The total heat stolen/given through each wall separating adjacent dwellings was calculated as per Equation (5).

$$Q_{stolen/given} = \sum_{h=1}^{8760} Q_h \tag{5}$$

In this way, the heat calculated as per Equations (4) and (5) is positive when the reference environment is gaining heat; on the contrary, it is negative when the heat is transferred from the reference environment to the adjacent one.

For each i-th apartment, the heat flowing through the walls adjacent with other apartments was calculated and it was assessed how much these heat flows would impact on the total energy consumption of the apartment itself as per Equation (6).

$$\frac{Q_{stolen/given,i}}{Q_{tot,i}}{\cdot}100 \tag{6}$$

The results are reported for each apartment in Table 4 for the base scenario first.

Table 4. Percentage * of heat given or stolen between adjacent apartments, base scenario.

from \ to	1C	2A	3B	4C	5A	6B	7C	8A	9B	Total
1C		44 (0.9%)		285 (6.2%)						329 kWh (7.1%)
2A	−44 (−0.7%)		2 (0.0%)		18 (0.3%)					−23 kWh (−0.4%)
3B		−2 (0.0%)				−46 (−0.6%)				−49 kWh (−0.7%)
4C	−285 (−4.6%)				−29 (−0.5%)		−323 (−5.2%)			−638 kWh (−10.2%)
5A		−18 (−0.4%)		29 (0.6%)		−13 (−0.3%)		−181 (−3.6%)		−183 kWh (−3.6%)
6B			46 (0.8%)		13 (0.2%)				148 (2.4%)	207 kWh (3.4%)
7C				323 (4.2%)				5 (0.1%)		328 kWh (4.3%)
8A					181 (2.7%)		−5 (−0.1%)		86 (1.3%)	261 kWh (3.9%)
9B						−148 (−1.8%)		−86 (−1.0%)		−234 kWh (−2.8%)

* to be read row by column, (-) signs meaning that the row apartment is giving energy to the column apartment; (+) signs meaning that the row apartment is stealing energy to the column apartment.

It is worth to observe that results obtained by this calculation depend both on the set-point temperature and the 'position' of the dwelling with respect to the others. For sake of completeness, a summary of the average seasonal value of the set-point temperature used to simulate the base scenario is given in Table 5.

Generally speaking, results show that in case of full operation of the building, the phenomenon of heat-thefts is almost negligible, representing on average 0.5% of the annual energy consumption for heating purposes of a single dwelling with the maximum absolute value of 6.2%. Observing the apartment with the lowest average set-point temperature (1C), it is possible to notice that this "steals heat" from both apartments 2A and 4C (respectively 1.0% and 6.2% of its total energy need for space heating).

Table 5. Average seasonal values of set-point temperature used for the simulation of the base scenario.

Dwelling Number	Set-Point Temperature, Average Seasonal Values [°C]
1C	18.6
2A	20.5
3B	21.4
4C	20.4
5A	19.3
6B	19.6
7C	19.6
8A	19.5
9B	22.1

On the other hand, even a central apartment that could benefit from its position (as for example apartment 4C) can nullify this advantage when using set-point temperatures higher than the one of adjacent apartments, losing up to 10% of its energy need.

However, an unexpected behavior can be highlighted for apartments 2A, 5A and 8A. Although, in fact, apartment 5A has set the lowest average set point temperature, it still gives heat to both adjacent apartments on the same vertical (2A and 8A). This is because set-point temperature is reached in the apartments only during the operating hours of the heating plant, while, for the rest of the day, the temperature inside the apartment 5A is always higher than that in 8A and 2A, due to its favorable position. This is evident in Figure 7, which reports the indoor temperature for the above-mentioned three apartments in a representative week of the heating season.

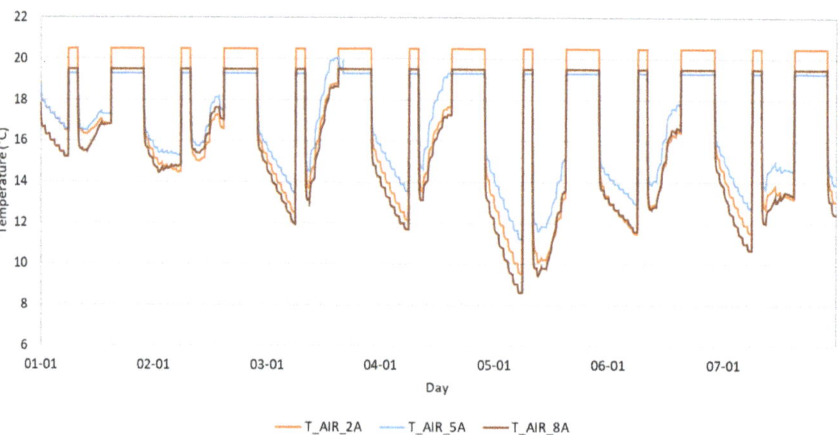

Figure 7. Simulated indoor temperature variation of apartments 2A, 5A and 8A in a representative week of the heating season, base scenario.

Referring to Figure 7, it is highlighted that in a day there are two "heating periods", according to the schedule of the heating plant (from 6 to 8 a.m. and from 15 to 22 p.m.). From the figure, it is evident that the installed heating system quickly manages to bring the apartment up to the set-point temperature, even in situations of daily temperature excursions. For the sake of completeness, Figure 8a,b show the trends of the daily indoor temperature in the above-mentioned apartments during the warm-up times simulated for January the 5th, which is representative of a cold winter day. From the figure, it can be highlighted that, even under high load conditions (i.e., low outdoor temperature), the heating system is very effective and set-point temperatures are reached in less than about 15 min. Generally speaking, this limited warm-up time mainly depends on the time constant of the heating system and, to a lesser extent, on that of the building. Obviously, this heating transient is not representative of the indoor comfort conditions of the room, that are determined by both the indoor air temperature and the mean radiant temperature, which depends on the thermal mass of the building.

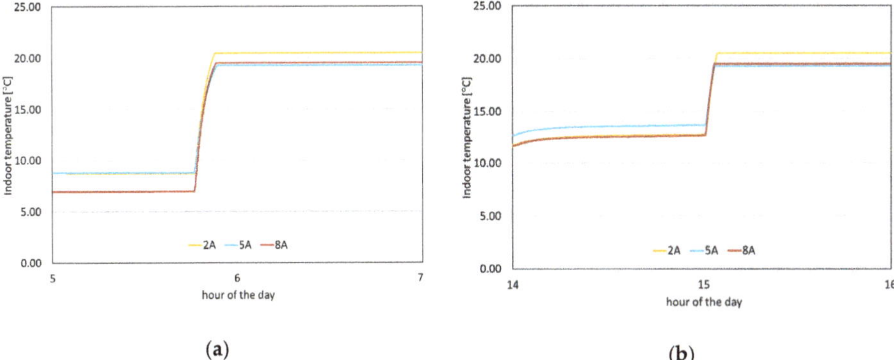

Figure 8. Simulated indoor temperature trends of apartments 2A, 5A and 8A during the warm-up time of a representative day: (**a**) from 5:00 a.m. to 7:00 a.m.; (**b**) from 14:00 p.m. to 16:00 p.m.

On the other hand, the low thermal mass of the building is the cause of high temperature drops when the generator is turned off, especially in apartments in disadvantaged positions, despite their heat thefts from adjacent dwellings. Finally, it is worth to observe the behavior of the building during non-heating hours with the related heat exchanges among apartments.

For the purpose of evaluating the heat gains and losses due to a change of operation, the authors simulated two different scenarios in addition to the base one: (i) scenario a, in which only apartment 5A was assumed to be unheated; (ii) scenario b, in which only apartment 2A was assumed to be unheated. In both cases, the set-point temperatures of all other apartments remained unchanged with respect to the base scenario. These were chosen as representative of the most and the least favored positions of block A (but similar considerations can be made for block B except for minor changes due to the different orientation): in fact, apartment 5A is located in the second floor and surrounded by heated spaces (apartments 2A, 6B, 4C and 8A) while apartment 2A is located at the first floor and its uninsulated floor is completely exposed to outdoor temperature.

Table 6 shows the results obtained from switching off the heating system of the dwelling 5A (scenario a).

Obviously, in this case all the thermal zones adjacent to apartment 5A give energy to it, with apartment 2A and 8A giving, respectively, 8.5% and 5.2% of their total energy need for space heating (in fact these apartments have the largest heat exchange surfaces). Overall, apartment 5A is able to steal 1249 kWh from its neighbors, which would have accounted for 25% of its energy consumption in the base scenario.

Referring to Figure 9, it is important to highlight that when the heating system is not working, in some hours of the week, apartment 5A still has its indoor air temperature higher than the one of both adjacent dwellings, confirming again the existence of an inversion in the direction of the heat flows already highlighted in the base scenario. In this scenario, the average seasonal difference of indoor temperature between apartment 5A and the other heated apartment is about 2.6 °C. Table 7 shows the results obtained from switching off the heating system of the dwelling 2A (scenario b) while Figure 10 shows the trend of the indoor temperature of the investigated apartments in a representative week of the heating season.

Table 6. Percentage * of heat given or stolen between adjacent apartments, scenario a, kWh and (%).

from \ to	1C	2A	3B	4C	5A	6B	7C	8A	9B	Total	
1C		25 (0.5%)		265 (5.7%)						290 (6.2%)	kWh
2A	−25 (−0.4%)		19 (0.3%)		−520 (−8.5%)					−526 (−8.6%)	kWh
3B		−19 (−0.3%)				−65 (−0.9%)				−85 (−1.2%)	kWh
4C	−265 (−4.2%)				−208 (−3.3%)		−303 (−4.7%)			−776 (−12.2%)	kWh
5A		520 (n.a.)		208 (n.a.)		156 (n.a.)		364 (n.a.)		1249 (n.a.)	kWh
6B		65 (1.1%)			−156 (−2.5%)				165 (2.6%)	74 (1.2%)	kWh
7C				303 (3.9%)				−13 (−0.2%)		290 (3.8%)	kWh
8A					−364 (−5.2%)		13 (0.2%)		101 (1.4%)	−250 (−3.6%)	kWh
9B					−165 (−1.9%)			−101 (−1.2%)		−266 (−3.1%)	kWh

* to be read row by column, (-) signs meaning that the row apartment is giving energy to the column apartment; (+) signs meaning that the row apartment is stealing energy to the column apartment.

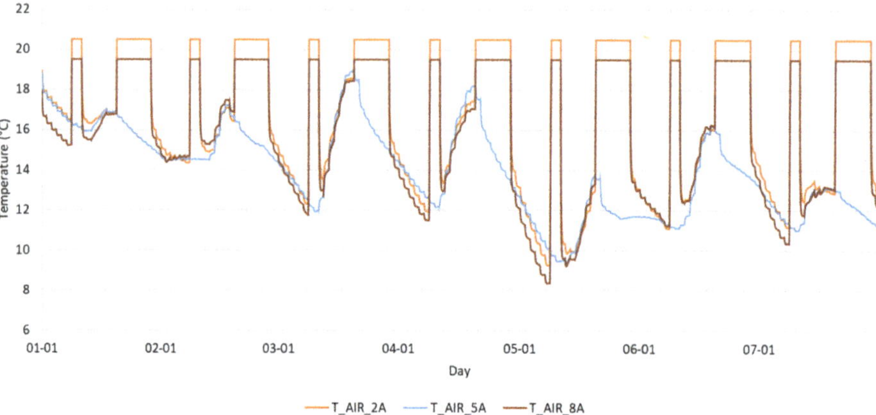

Figure 9. Indoor temperature variation of apartments 2A, 5A and 8A in a representative week of the heating season, scenario a.

The results show that apartment 2A is always able to gain heat from its neighbors (apartments 5A, 1C, 3B,) since its indoor temperature is always lower than that of the other ones, stealing up to 13.6% of the total energy need for space heating of apartment 5A. Overall, apartment 2A is able to steal 1160 kWh from its neighbors, which would have accounted for 20% of its energy consumption in the base scenario. However, in this scenario, the average seasonal difference of indoor temperature between apartment 2A and the other heated apartment is about 3.6 °C, which is 1.0 °C lower than the one of apartment 5A in the similar scenario.

From the analysis of the results it can be observed that heat transfer between a heated dwelling toward an adjacent unheated, ranges 20–25% of its theoretical energy need, which is much lower than the estimated 70–90% in available literature [10,12,13]. This is essentially due to the fact that the investigated building is poorly insulated (i.e., low thermal insulation both in the internal partitioning walls and towards the external environment); thus, the energy need of dwellings is mainly determined by their thermal dispersion towards the outdoor environment (compared to which the heat transfers toward other apartments are certainly lower). On the contrary, in [8–10,12,13] the investigated case-study buildings presented highly insulated external facades and they were mainly located in cold continental climates. The milder climatic conditions under which the present simulation has been carried out also affect the extent of heat stolen, thanks to not-negligible solar heat gains, which in scenario a determine an inversion of heat exchanges during the central hours of sunny days.

Table 7. Share * of heat given or stolen between adjacent apartments, scenario b, kWh and (%).

from \ to	1C	2A	3B	4C	5A	6B	7C	8A	9B	Total
1C		−193 (−4.0%)		315 (6.6%)						122 kWh (2.6%)
2A	193 (n.a.)		228 (n.a.)		739 (n.a.)					1160 kWh (n.a.)
3B		−228 (−3.1%)				−21 (−0.3%)				−249 kWh (−3.4%)
4C	−315 (−5.0%)				−54 (−0.9%)		−317 (−5.0%)			−686 kWh (−10.9%)
5A		−739 (−13.6%)		54 (1.0%)		10 (0.2%)		−107 (−2.0%)		−782 kWh (−14.5%)
6B			21 (0.3%)		−10 (−0.2%)				153 (2.5%)	164 kWh (2.7%)
7C			317 (4.1%)					2 (0.0%)		319 kWh (4.1%)
8A					107 (1.6%)		−2 (0.0%)		87 (1.3%)	193 kWh (2.8%)
9B						−153 (−1.8%)		−87 (−1.0%)		−240 kWh (−2.8%)

* to be read row by column, (-) signs meaning that the row apartment is giving energy to the column apartment; (+) signs meaning that the row apartment is stealing energy to the column apartment.

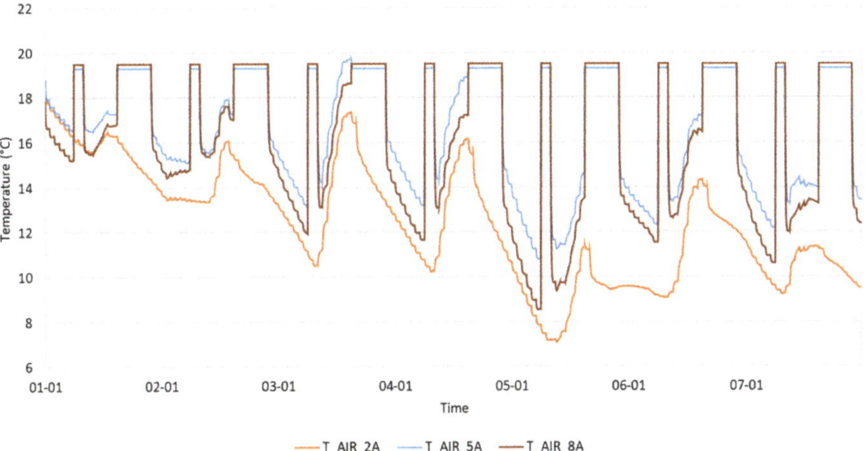

Figure 10. Indoor temperature variation of apartments 2A, 5A and 8A in a representative week of the heating season, scenario b.

As a matter of fact, in a Mediterranean climate, the compensation of heat costs based on average indoor temperature [7–9] or on static heat flows [11], would tend to overestimate the stolen heat between apartments, not taking into consideration the dynamic behavior of this phenomenon. It is highlighted that the results obtained within the present research could be useful to analyze similar buildings in terms of climatic conditions and orientation. However, in order to allow a greater and general applicability of the results, it would be necessary to carry out a dedicated sensitivity analysis to the variation of the abovementioned parameters.

4. Conclusions

In this work the issue of heat thefts between adjacent apartments has been investigated by performing a dynamic simulation on a case-study building. To this end, a dynamic model of a real social housing building supplied by a centralized heating plant has been implemented using TRNSYS simulation software, validated with real energy consumption data and calibrated in order to obtain a maximum deviation from the measured data in a range of +6/−12% per apartment. The final estimation error of the model on the total building energy consumption was −2%.

Three scenarios were simulated, in order to highlight the differences between the possible operation of some apartments: a base scenario considering a full operation of the heating plant (i.e., all apartments heated at different set-point temperatures) and two additional scenarios in which two apartments chosen as representative of favorable and unfavorable positions were considered, individually, unheated.

The analysis highlighted that the amount of the heat thefts due to different set-point temperatures in case of full operation of the building is almost negligible in low-insulated buildings, such as the analyzed one. In particular, the share of heat exchanges calculated with respect to the total annual energy need for space heating of single apartments varies between 0 and 6.2%. Dynamic effects of inversion of the heat transfers between adjacent apartments have also been highlighted in the same scenario when an intermediate floor apartment, although having a lower set-point temperature during the heating hours, registered a higher indoor temperature compared to when the heating system was not working.

It has also been highlighted that, for the same end-user behavior (i.e., switching off the heating system), the benefit obtained from the favorably positioned apartment is higher to that of the apartment located in a disadvantaged position in terms of indoor temperature, although the amount of heat thefts of the apartment with a disadvantaged position is higher.

In fact, depending on apartment location, the thermal mass of the apartments in a favorable position would prevent their indoor temperatures from falling below acceptable threshold values if the users of those dwellings decide to turn down their radiators through the TRVs. On the contrary, this would not be true for apartments in unfavorable positions.

The results obtained show that greater attention should be paid to heat thefts for the purpose of heat cost allocation, especially in buildings with occasional and uneven occupation and/or operation and in buildings with low thermal performances (in particular those with poorly insulated partition surfaces between apartments).

More attention should be given to the heat theft phenomenon both in terms of: (*i*) design of new buildings or of retrofit interventions. The building envelope and the heating system should be designed keeping in mind the real operating conditions while normally, in design phase, the apartments are considered to be always heated, even in buildings (such as mountain residences) normally employed for occasional occupation; (*ii*) development of specific compensation techniques for heat cost allocation in buildings with particular reference to the estimation of the "involuntary" part of energy consumption (i.e., the one which is not determined by the end-user behavior). Moreover, this study highlights the need of an extensive analysis regarding compensation due to heat transfers between adjacent apartments in mild climatic conditions, which is still required to obtain a heat cost allocation methodology easily replicable for designers and building managers.

Author Contributions: Conceptualization M.D., G.P.; methodology, M.D., G.F., L.C.; formal analysis, L.C., V.B.; investigation, L.C., V.B.; resources, M.D., G.P.; data curation, L.C., V.B.; writing—original draft preparation, L.C.; writing—review and editing, L.C., G.F., M.D., V.B., G.P.; supervision, G.F., M.D., G.P. All authors have read and agreed to the published version of the manuscript.

Funding: This work has been developed under the projects "Ricerca di Sistema Elettrico PAR 2016" funded by ENEA (grant number I12F16000180001) and "PRIN Riqualificazione del parco edilizio esistente in ottica NZEB" funded by MIUR (grant number 2015S7E247_002). The authors wish to thank ATER of Frosinone, the Territorial Agency for Social Housing, for the technical support during the on field experimental campaign.

Conflicts of Interest: The authors declare no conflict of interest

References

1. European Union. Energy Roadmap. 2050. Available online: https://ec.europa.eu/energy/sites/ener/files/documents/2012_energy_roadmap_2050_en_0.pdf (accessed on 1 February 2020).
2. International Energy Agency. Energy Efficiency: Buildings. The Global Exchange for Energy Efficiency Policies, Data and Analysis. Available online: https://www.iea.org/topics/energyefficiency/buildings/ (accessed on 1 August 2019).
3. European Commission. *Guidance Note on Directive 2012/27/EU on Energy Efficiency, Amending Directives 2009/125/EC and 2010/30/EC, and Repealing Directives 2004/8/EC and 2006/32/EC Article 7: Energy Efficiency Obligation Schemes Accompanying the Document Communication from the Commission to the European Parliament and the Council Implementing the Energy Efficiency Directive—Commission Guidance*; European Commission: Brussels, Belgium, 2013.
4. European Commission. *Directive (EU) 2018/844 of the European Parliament and of the Council of 30 May 2018 Amending Directive 2010/31/EU on the Energy Performance of Buildings and Directive 2012/27/EU on Energy Efficiency*; European Commission: Brussels, Belgium, 2018.
5. Canale, L.; Dell'Isola, M.; Ficco, G.; Cholewa, T.; Siggelsten, S.; Balen, I. A comprehensive review on heat accounting and cost allocation in residential buildings in EU. *Energy Build.* **2019**, *202*. [CrossRef]
6. Dell'Isola, M.; Ficco, G.; Canale, L.; Frattolillo, A.; Bertini, I. A new heat cost allocation method for social housing. *Energy Build.* **2018**, *172*, 67–77. [CrossRef]
7. Liu, L.B.; Fu, L.; Jiang, Y.; Guo, S. Major issues and solutions in the heat-metering reform in China. *Renew. Sustain. Energy Rev.* **2011**, *15*, 673–680. [CrossRef]
8. Siggelsten, S. Reallocation of heating costs due to heat transfer between adjacent apartments. *Energy Build.* **2014**, *75*, 256–263. [CrossRef]
9. Michnikowski, P. Allocation of heating costs with consideration to energy transfer from adjacent apartments. *Energy Build.* **2017**, *139*, 224–231. [CrossRef]
10. Gafsi, A.; Lefebvre, G. Stolen heating or cooling energy evaluation in collective buildings using model inversion techniques. *Energy Build.* **2003**, *35*, 293–303. [CrossRef]
11. Pakanen, J.; Karjalainen, S. Estimating static heat flows in buildings for energy allocation systems. *Energy Build.* **2006**, *38*, 1044–1052. [CrossRef]
12. Lukić, N.; Nikolić, N.; Timotijević, S.; Tasić, S. Influence of an unheated apartment on the heating consumption of residential building considering current regulations—Case of Serbia. *Energy Build.* **2017**, *155*, 16–24. [CrossRef]
13. Xue, P.; Yang, F.; Zhang, Y.; Zhao, M.; Xie, J.; Liu, J. Quantitative study on adjacent room heat transfer: Heating load and influencing factors. *Sustain. Cities Soc.* **2019**, *51*, 101720. [CrossRef]
14. Siggelsten, S. Heat cost allocation in energy efficient multi-apartment buildings. *Cogent Eng.* **2018**, *5*. [CrossRef]
15. Italian National Unification. *UNI 10200. Impianti Termici Centralizzati di Climatizzazione Invernale ed Acqua Calda Sanitaria—Criteri di Ripartizione Delle Spese di Climatizzazione ed Acqua Calda Sanitaria (in Italian Language Only)*; UNI: Milan, Italy, 2018.
16. European Commission. *Directive 2012/27/EU of the European Parliament and of the Council of 25 October 2012 on Energy Efficiency, Amending Directives 2009/125/EC and 2010/30/EU and Repealing Directives 2004/8/EC and 2006/32/EC Text with EEA Relevance*; European Commission: Brussels, Belgium, 2012.
17. Italian Republic. *Decreto Legislativo 4 Luglio 2014, n. 102 Attuazione Della Direttiva 2012/27/UE Sull'efficienza Energetica, che Modifica le Direttive 2009/125/CE e 2010/30/UE e Abroga le Direttive 2004/8/CE e 2006/32/CE (in Italian Language Only)*; Gazzetta Ufficiale della Repubblica Italiana: Rome, Italy, 2014.
18. European Committee for Standardization (CEN). *EN 442-1: Radiators and Convectors—Part 1: Technical Specification and Requirements*; CEN: Brussels, Belgium, 2014.
19. European Committee for Standardization (CEN). *EN 442-2: Radiators and Convectors—Part 2: Test Methods and Rating*; CEN: Brussels, Belgium, 2014.
20. Dell'Isola, M.; Ficco, G.; Canale, L.; Palella, B.I.; Puglisi, G. An IoT Integrated Tool to Enhance User Awareness on Energy Consumption in Residential Buildings. *Atmosphere* **2019**, *10*, 743. [CrossRef]
21. University of Wisconsin. TRNSYS 17: A transien system simulation program. *Sol. Energy Lab.* 2012. Available online: http://www.trnsys.com/ (accessed on 19 July 2019).

22. Attia, S.; Beltrán, L.; Herde, A.D.; Hensen, J. Architect friendly: A comparison of ten different building performance simulation tools. In Proceedings of the 11th International Building Performance Simulation Association Conference and Exhibition 2009, Glasgow, Scotland, 27–30 July 2009; pp. 204–211.
23. Crawley, D.B.; Hand, J.W.; Kummert, M.; Griffith, B.T. Contrasting the capabilities of building energy performance simulation programs. *Build. Environ.* **2006**, *43*. [CrossRef]
24. University of Wisconsin-Madison. TRNSYS 17 documentation, Volume 5, Multi zone building modeling with type56 and TRNBuild. *Sol. Energy Lab.* **2012**.
25. Ente Nazionale Italiano di Unificazione. *UNI 11300-1 Prestazioni Energetiche Degli Edifici—Parte 1: Determinazione del Fabbisogno di Energia Termica Dell'edificio per la Climatizzazione Estiva ed Invernale (in Italian)*; Ente Nazionale Italiano di Unificazione: Milano, Italy, 2014.
26. Italian Republic. *Decreto del Presidente della Repubblica 26 Agosto 1993, n. 412 Regolamento Recante Norme per la Progettazione, L'installazione, L'esercizio e la Manutenzione Degli Impianti Termici Degli Edifici ai Fini del Contenimento dei Consumi di Energia, in Attuazione Dell'art. 4, Comma 4, Della Legge 9 Gennaio 1991, n. 10. (GU Serie Generale n.242 del 14-10-1993—Suppl. Ordinario n. 96)*; Gazzetta Ufficiale della Repubblica Italiana: Rome, Italy, 1993.
27. Arpino, F.; Cortellessa, G.; Dell'Isola, M.; Ficco, G.; Marchesi, R.; Tarini, C. Influence of installation conditions on heating bodies thermal output: Preliminary experimental results. *Enrgy Procedia* **2016**, *101*, 74–80. [CrossRef]

Article

Experimental Evaluation and Numerical Simulation of the Thermal Performance of a Green Roof

Claudia Guattari, Luca Evangelisti, Francesco Asdrubali * and Roberto De Lieto Vollaro

Department of Engineering, Roma TRE University, Via Vito Volterra 62, 00146 Rome, Italy;
claudia.guattari@uniroma3.it (C.G.); luca.evangelisti@uniroma3.it (L.E.);
roberto.delietovollaro@uniroma3.it (R.D.L.V.)
* Correspondence: francesco.asdrubali@uniroma3.it

Received: 5 February 2020; Accepted: 28 February 2020; Published: 4 March 2020

Abstract: In the building sector, both passive and active systems are essential for achieving a high-energy performance. Considering passive solutions, green roofs represent a sustainable answer, allowing buildings to reach energy savings, and also reducing the collateral effect of the Urban Heat Island (UHI) phenomenon. In this study, a roof-lawn system was investigated by means of an extended measurement campaign, monitoring the heat transfer across the roof. Heat-flow meters and air- and surface-temperature probes were applied in a real building, in order to compare the performance of the roof-lawn system with a conventional roof. This experimental approach was followed to quantify the different thermal behaviors of the building components. Moreover, an equivalent thermal model of the roof-lawn system was studied, in order to obtain the equivalent thermal properties of the roof, useful for setting building models for yearly energy simulations. The roof-lawn system revealed its advantages, showing a higher thermal inertia with no overheating in summertime and a lower thermal transmittance with energy savings in wintertime, and, consequently, better indoor conditions for the occupants of the building.

Keywords: green roof; experimental investigation; thermal performance; measurements; simulations

---

1. Introduction

In recent years, urban areas' growth and the consequent sources of pollution have led to an increase in terms of global warming. In addition, the so-called Urban Heat Island (UHI) phenomenon has developed. UHI is represented by the temperature rises in areas characterized by a high urban fabric, if associated to the surrounding rural areas [1]. The evaluation of the UHI phenomenon is crucial to design plants effectively and to evaluate the buildings' energy needs. It is noteworthy to suggest interventions for the mitigation of this phenomenon, lowering its intensity. Achieving this goal is fundamental to reducing the increasing building energy consumption, especially during the warmer months [2,3]. If, on the one hand, the internal temperatures of the buildings can be set lower, the same cannot be done for the external environment, with the exception of countermeasures finalized for the reduction in UHI effects. Furthermore, the high temperatures that occur in cities during the warmer months can involve substantial and damaging effects on daily life [4–7].

Therefore, the UHI phenomenon and its countermeasures are topics of interest in the scientific literature. Several works aimed at assessing UHI intensity suggest strategies to reduce its impacts, such as building design approaches [8–10]. Several mitigation measures were recommended in order to mitigate UHI impacts on environmental, energy, economic and social aspects. Some of them are based on the existing correlation between the UHI and the polluting gases in the atmosphere. Consequently, the reduction in pollutant emissions has a direct influence on the UHI, and the UHI reduction has a direct influence on building energy consumption [11,12]. All this can be achieved by means of a rational buildings project, but also considering the emissions reduction induced by transport or industrial sources. Further mitigation solutions can be reached by employing construction materials with a high albedo.

This research focused on the assessment of the effectiveness of a green roof thermal behavior. The goal was to analyze the thermophysical properties and thermal behavior to estimate the real effect on the building's energy performance optimization.

Several measuring instruments were installed for observing heat transfer phenomena across the green roof. The experimental campaign consisted of acquiring data from both the green roof and a nearby conventional roof, in order to compare the thermal performance.

Moreover, the roof-lawn system, characterized by a multilayer structure, was reproduced using a Finite Element Model (FEM) code for obtaining equivalent thermophysical properties. It is worthy to mention that the compositions of the roofs were known but the thermophysical properties of each layer were undetermined. In addition, the green roof is made of five layers, of which the roof-lawn is non-homogeneous, and consists of a layer of grass and the underlying substrate. On the other hand, the original roof is made by only three layers (explained in the following section), characterized by conventional materials. This approach was applied for achieving suitable information for creating building energy models for the simulation of yearly energy needs.

The roof-lawn system showed its advantages, pointing out a higher thermal inertia with no overheating during the warmer months and a lower thermal transmittance. This resulted in lower energy demands and, consequently, better internal environmental conditions.

Consequently, it is possible to assert that a green roof can enhance the thermal inertia of a roof, thus increasing the internal comfort, reducing energy needs [13–19]. In addition, if installed in high-density urban areas, a green roof can offer positive contribution against the UHI phenomenon, also absorbing polluting gases.

The novelty of this work is related to two different aspects: on one hand, a yearly monitoring of a green roof can be useful for readers, showing the thermal behavior of a roof-lawn resting on an existing roof. This is a long first analysis of a deeper optimization study of the structural part of the roof, aimed at designing a stratigraphy better able to work with the overlying roof-lawn. On the other hand, an inverse method was applied here to evaluate the equivalent thermophysical properties of a green roof inside an innovative methodological approach.

2. Materials and Methods

2.1. On-Site Measurements

The experimental survey was done by analyzing a roof-lawn system installed on the roof of a single-story building (see Figure 1a) situated in the countryside near the city of Latina (about 70 km south of Rome). This research aimed at evaluating the thermal characteristics of a green roof by means of an extended measurement campaign. The roof-lawn system is an innovative patent, characterized by the species of the Zoysia genus (distinguished by a slow growth and a typical wave effect). The realization of the water system only reintegrates the losses due to evapotranspiration. The level of maintenance required for the green roof is extremely low, since the routine maintenance is not required. Once it reached the maximum growth, the vegetation has a characteristic wave effect with average foliage heights that do not exceed 25 cm. The whole system components are reported in Figure 1b.

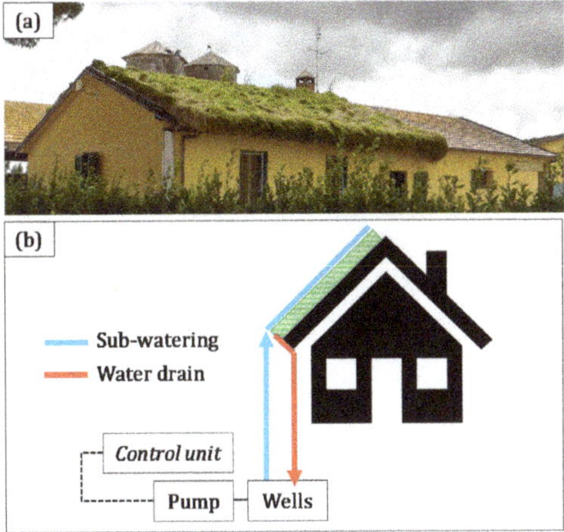

Figure 1. (**a**) Single-story building on which the green roof is installed and (**b**) roof-lawn whole system (an elaboration from [20]).

The stratigraphy of the roof is shown in Figure 2. The green roof is characterized by different layers: on the upper part of the roof, a waterproofing sheath was installed to avoid water infiltration; over that, the green roof was built on a draining mat and an inorganic substrate. The structural part of the roof is composed of a reinforced concrete layer, with a thickness of about 8 cm. The overall thickness of the green roof is equal to 20 cm.

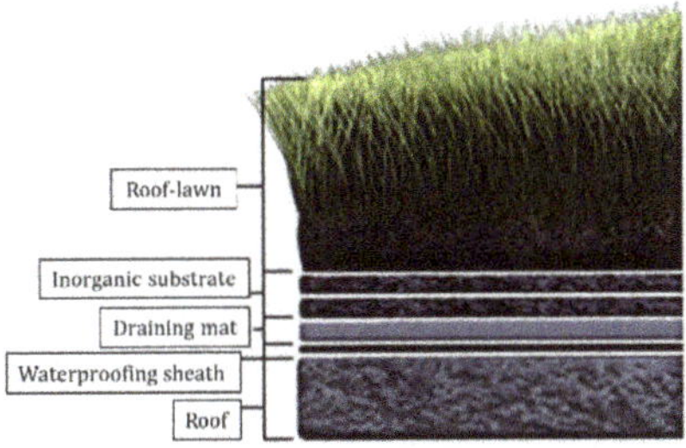

Figure 2. Green roof stratigraphy.

Only a half of the building roof is characterized by the installation of the roof-lawn system, while the other half remained in the previous condition. The original roof is made of a reinforced concrete slab with a thickness of about 8 cm, covered with tiles. A waterproof membrane is installed on the concrete slab. The two rooms under the roofs are characterized by the same orientation and the same occupation rate.

To assess the behavior of the green roof and define the characteristics of the system in terms of stationary and dynamic thermal performance, heat flow sensors, air temperatures probes and surface temperature sensors were installed, as shown in Figure 3. The measuring instruments' technical data are reported in Table 1.

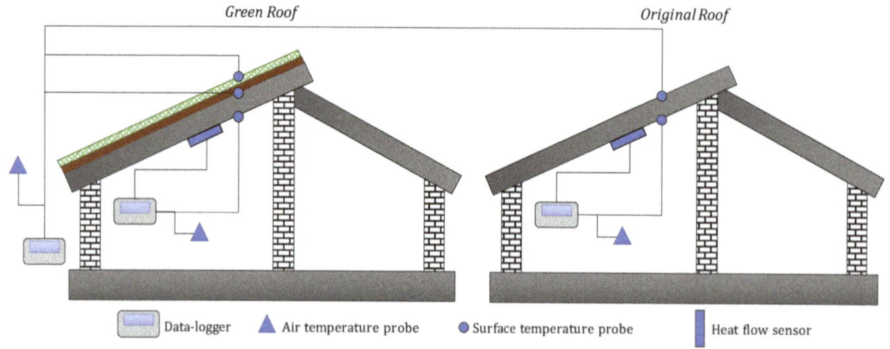

Figure 3. Measurement apparatus schema.

Table 1. Measuring instruments technical data.

Measuring Instrument	Manufacturer	Model	Measuring Range	Resolution	Accuracy
Heat-flow sensor	Hukseflux	HFP01	$-2000 \div 2000$ W/m^2	0.01 W/m^2	5% on 12 h
Thermometer	LSI	Pt100	$-40 \div 80$ °C	0.01 °C	0.10 °C (0 °C)
Surface temperature sensor	LSI	EST124	$-40 \div 80$ °C	0.01 °C	0.15 °C (0 °C)

It is worth mentioning that the external surface temperature probe installed on the upper layer of the green roof was inserted below a first small layer of soil, in order to guarantee the best thermal contact. The schematic representation reported in Figure 3 could be misleading, as it represents the external surface temperature probe (blue circle) on the outermost part of the green layer. Of course, the upper part of a green roof is made of grass and surface temperatures cannot be measured. Therefore, the surface temperature probe was installed, placing the sensor in the upper part of the soil, where the grass grows.

All sensors were connected to the data-loggers, recording heat fluxes, air and surface temperatures with a timestep equal to 10 min, for 24 h per day. The measurement campaign started in October 2018 and finished after one year, in September 2019. All the acquired data were used to calculate the thermal transmittances (also known as U-value or merely U) of the green roof and the original roof.

Heat fluxes and air temperatures can be used for calculating the U-value by applying the following formula

$$q = U(T_i - T_e) \tag{1}$$

where q is the heat flux density, and T_i and T_e are the air temperature in the internal and external environment, respectively. According to the standard ISO 9869-1 [21], heat fluxes and air temperature values were used for calculating the stationary U-value of the roof by applying the average progressive method, following the formula

$$U = \frac{\sum_{j=1}^{N} q_j}{\sum_{j=1}^{N} \left(T_{ij} - T_{ej}\right)} \tag{2}$$

where N is the total registered samples.

In addition, using surface temperatures instead of air temperatures, the thermal conductance (C-value or only C) of the roof can be deduced, by applying the following equation

$$C = \frac{\sum_{j=1}^{N} q_j}{\sum_{j=1}^{N} \left(T_{sij} - T_{sej}\right)} \tag{3}$$

where T_{si} and T_{se} are the internal and external surface temperatures, respectively.

Moreover, internal and external surface temperatures were also used to obtain information about the dynamic thermal performance of the roofs, in terms of heat waves' phase shift and decrement factor.

The heat waves' Phase Shift (PS) can be evaluated as the time difference between the maximum value of the internal surface temperature and the maximum value of the external surface temperature of the roof [22]

$$PS = t_{T_{si}^{MAX}} - t_{T_{se}^{MAX}} \tag{4}$$

The Decrement Factor (DF) can be defined as follows [22]

$$DF = \frac{T_{si}^{MAX} - T_{si}^{MIN}}{T_{se}^{MAX} - T_{se}^{MIN}} \tag{5}$$

where T_{si}^{MAX} and T_{si}^{MIN} are the maximum and the minimum internal surface temperatures registered in a day, and, in turn, T_{se}^{MAX} and T_{se}^{MIN} are the maximum and the minimum surface external temperatures registered in a day.

2.2. Methodology

In order to assess the thermal behavior of the green roof and its effectiveness in terms of energy savings, it was necessary to analyze and compare the thermal performance of the two parts of the roof. It is worth to mentioning that the stratigraphies of the two roofs were known but the thermophysical properties of each layer were undetermined. In addition, the green roof is made of five layers, of which the roof-lawn is non-homogeneous, and it consists of a layer of grass and the underlying soil. On the other hand, the original roof is made of only three layers, characterized by conventional materials. For this reason, data obtained from the on-site measurement campaign were used for generating a model through Comsol Multyphysics software [23,24]. The experimental system registered data useful in setting the boundary conditions in the model. Taking into account the direction of the heat flux caused by the difference in temperature between the two sides of the roof, a thermal input, equal to the experimental surface temperature trend, was set. The internal measured air temperatures and the heat fluxes were employed to calculate proper internal heat transfer coefficients.

The stratigraphy of the actual green roof was reproduced in Comsol as a single homogeneous layer, characterized by equivalent thermophysical properties, following the method demonstrated in [25]. Simulations were performed in different periods of the year, all characterized by no thermal inversion between internal and external air temperatures and heat fluxes characterized by the same direction for a better calculation of the internal heat transfer coefficients (h_{int}) [26].

On-site measurements were used as boundary conditions in the model: external surface temperature values were used as an external forcing function; on the other side of the equivalent layer, a heat transfer based on the equation $q = h_{int}(T - T_{env})$ was set, where T is the internal surface temperature and T_{env} is the temperature of the environment, outside the simulated domain (in this case corresponding to the indoor air temperature). From experimental measurements, h_{int} values along time were calculated and used in the simulation code. Thus, different equivalent thermophysical properties were iteratively tested and the internal surface temperatures were simulated. The search for equivalent thermophysical properties aimed to obtain the best reproduction of the behavior of the green roof, to obtain the best correspondence between the measured and simulated internal temperatures of the surface. The desired condition to stop searching for the best parameters is represented by a Model Efficiency (EF) value greater than 0.9 [23], expressed as

$$EF = \frac{\sum_{i=1}^{N}(m_i - \overline{m})^2 - \sum_{i=1}^{N}(s_i - m_i)^2}{\sum_{i=1}^{N}(m_i - \overline{m})^2} \tag{6}$$

where m_i is the measured value at time t_i, s_i is the simulated value for each time t_i, $\overline{m}$ is the average of the measured values and N is the total number of samples. EF can understand the capability of the equivalent structure to reproduce the original one's behavior, showing values between 0 and 1 (which indicates that measured and simulated data are equal).

Finally, the thermophysical properties obtained by the green and the original roofs were used in the energy simulation software TRNSYS in order to obtain the annual energy needs of a detached building. Thus, the comparison between the energy demands allowed the evaluation of the advantages deriving from the installation of a green roof. Considering that the solar reflectance of green roofs varies between 0.3 and 0.5 depending on the plant types [27], here, a reflectivity of 0.3 (typical value for leaves) was used in the model. The detached building was modelled considering walls consisting of a 0.22 m layer of concrete and a 0.04 m XPS layer, plastered on both sides, with a U-value of 0.600 W/(m^{2}K). The windows (U-value of 5.61 W/(m^{2}K)) are characterized by a total area equal to 18 m^2. The walls' solar absorptance coefficient was set equal to 0.6. The infiltration rate was set at 0.3 1/h and the indoor set-point temperatures for heating and cooling were set as equal to 20 and 26 °C, respectively. The flow-chart of the applied methodological approach is reported in Figure 4.

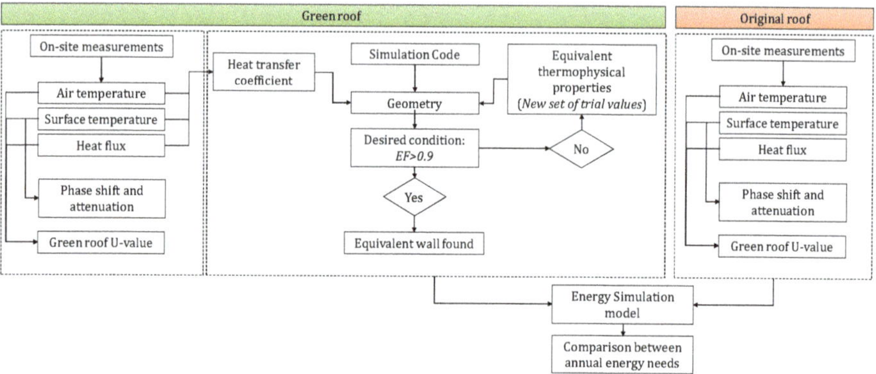

Figure 4. Methodological approach flow-chart.

3. Results and Discussion

3.1. Experimental Investigation

The results obtained during the measurement campaign are reported in this section. The monitoring was conducted from October 2018 to September 2019. The results of one year of measurements are a wealth of collected data. For the sake of brevity, partial data referring to winter and summer are presented here. This approach can be useful for readers to understand the thermal behavior of the green roof under different climatic conditions.

Regarding the winter season, the time range chosen to be analyzed was from 1st of December to 31st of December. This period was selected due to the low temperatures recorded. During this month the outdoor temperature range was between 3.83 and 14.16 °C.

The data registered during this period are reported in Figure 5, where Figure 5a refers to the green roof and Figure 5b to the traditional roof (distinguished in the label as "*ORIGINAL*"). In Figure 5a, the heat flux (called *q_GREEN*) is represented by the continuous red line, the indoor air temperature (called *Ti_GREEN*) is represented by the dashed green line and the internal surface temperature (called *Tsi_GREEN*) is represented by the continuous green line. On the other side of the roof, the surface temperature registered under the green roof was called *Textrados_GREEN* (orange dotted line) and the temperature of the upper layer of the green roof (measured inserting the sensor below a first small layer of soil) was called *Ts_roof-lawn* (continuous line of light green color).

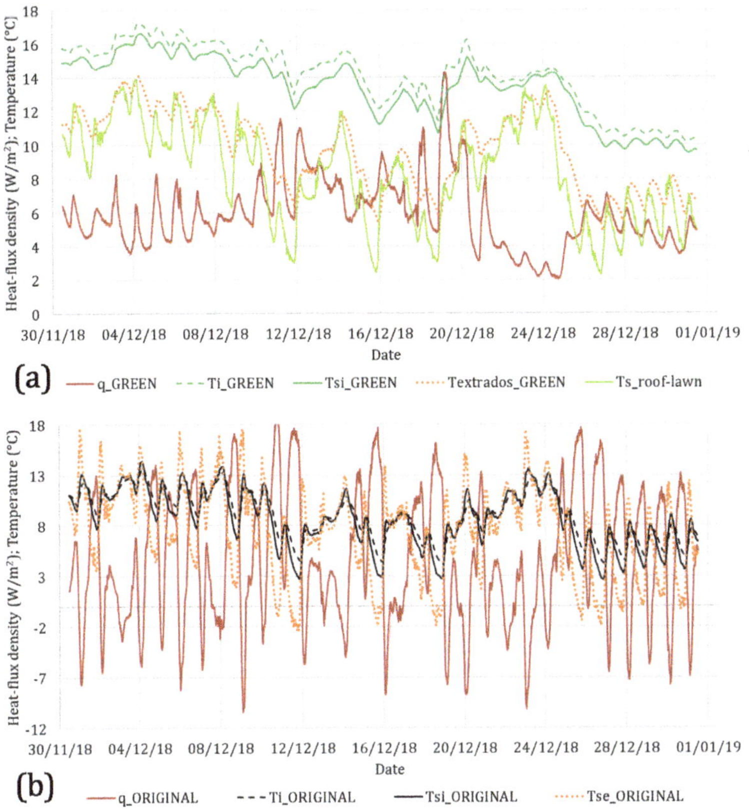

Figure 5. Heat fluxes, indoor temperatures and internal surface temperatures registered during winter for the green roof (**a**) and the original one (**b**).

On the other hand, in Figure 5b, the heat flux (called *q_ORIGINAL*) is depicted by the continuous red line, the indoor air temperature (called *Ti_ORIGINAL*) is represented by the dashed black line and the internal surface temperature (called *Tsi_ORIGINAL*) is depicted by the continuous black line. On the other side of the roof, the external surface temperature was called *Tse_ORIGINAL* and it is represented by the orange dotted line.

Observing both heat fluxes and temperatures, it is possible to notice a more stable thermal behavior of the green roof if compared with the original one. Referring to the green roof, analyzing the indoor temperatures and the internal surface temperatures, it is possible to obtain average values equal to 14.14 and 13.39 °C, respectively. On the contrary, for the original roof, an average indoor temperature of 8.95 °C and an average internal surface temperature of 8.58 °C can be observed. In order to provide a straightforward comparison among the acquired heat fluxes and temperatures, the upper limit of the ordinate axes was limited to the same value. Considering the green roof, the highest value for the heat flux was equal to about 14 W/m^2, a much lower value than those registered for the original roof. The different inertial behavior of the original roof allowed the observation of a strong fluctuation of heat flows, also showing negative values. Taking into account the external surface temperatures, it is possible to notice that the original roof is characterized by higher values. The main reason to explain these results is strictly related to the thermophysical characteristics of the employed materials. The roof covering tiles have a high solar radiation absorption coefficient, unlike green roofs, which are characterized by evapotranspiration phenomena.

The registered data allowed us to calculate the steady-state thermal transmittance of the roofs. Applying Equation (2), the U-values shown in Figure 6 were computed; it can be noticed that, after a few days, both thermal transmittances tend to reach a stationary value. A thermal transmittance of 1.361 W/(m^{2}K) was obtained for the green roof. On the other hand, a value of 3.021 W/(m^{2}K) was found for the original roof. Therefore, comparing the green and the original roofs U-values, a percentage difference of about −55% can be highlighted.

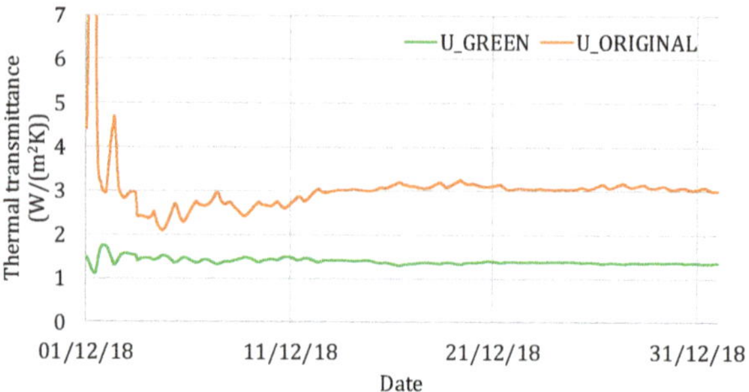

Figure 6. Green roof and original roof U-values.

Regarding the middle season, the time range chosen to be analyzed was from 10th of June to 26th of June. During this period, the outdoor temperatures range was between 5.53 and 26.23 °C. The heat fluxes, indoor temperatures and internal surface temperatures recorded during middle season are reported in Figure 7, where Figure 7a refers to the green roof and Figure 7b to the original roof. The graphical representation of heat fluxes and temperatures is the same as that used in Figure 5. In order to provide a straightforward comparison among the acquired heat fluxes, the upper and lower limits of the secondary axes are the same. Even if the average indoor temperatures and the average internal surface temperatures are not significantly different between the two roof configurations, during this period it is possible to notice a more stable thermal behavior of the green roof. As a matter of fact, it is possible to notice that the heat fluxes along time of the original roof show much higher fluctuations than those observed for the green roof (the heat flux of the original roof ranged between −32.56 and 19.04 W/m^2). Due to the thermophysical properties of the material used for the roof, in this case the external surface temperatures showed different values. The external surface temperatures of the original roof reached values above 40 °C, while the external temperatures of the roof-lawn system reached values below 22 °C.

Regarding the summer season, the time range chosen to be analyzed was from 10th of June to 26th of June. This period was selected due to the high temperatures recorded. During this period, the outdoor temperatures range was between 16.0 and 33.4 °C.

The heat fluxes, indoor temperatures and internal surface temperatures recorded during summer are reported in Figure 8, where Figure 8a refers to the green roof and Figure 8b to the original roof. In this case, the graphical representation of heat fluxes and temperatures is the same as that used in Figure 5. In order to provide a straightforward comparison among the registered heat fluxes, the upper and lower limits of the secondary axes are the same. It is possible to observe a more stable thermal behavior of the green roof also during summertime. Considering the green roof, the average indoor temperatures and the average internal surface temperatures are equal to 24.80 and 24.60 °C, respectively.

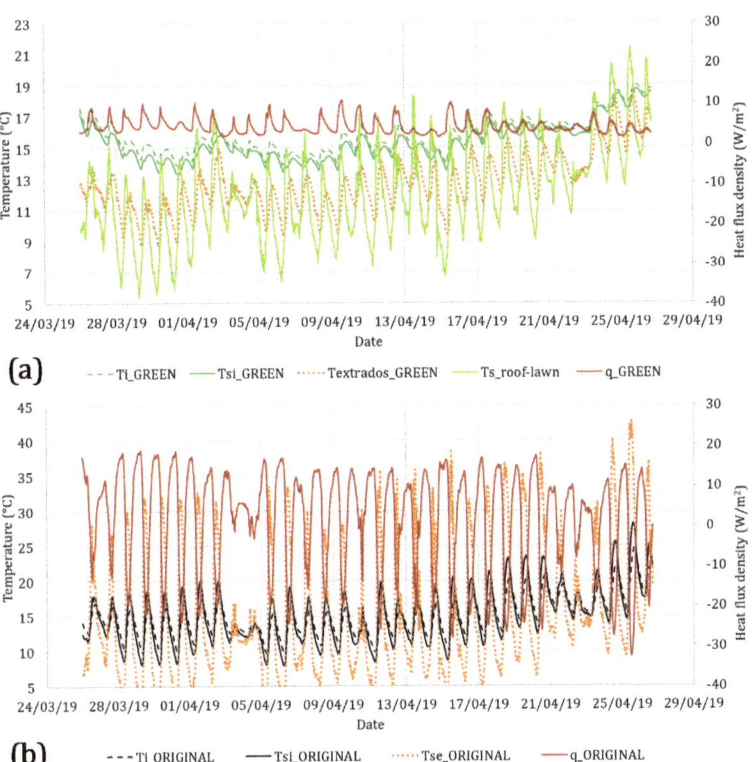

(a)

(b)

Figure 7. Heat fluxes, indoor temperatures and internal surface temperatures registered during middle season for the green roof (**a**) and the original one (**b**).

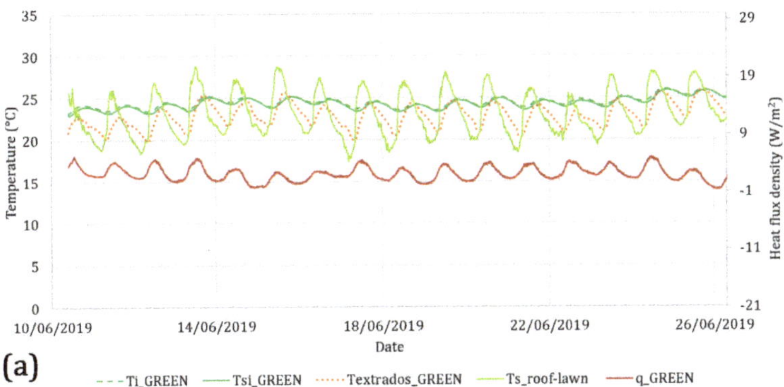

(a)

Figure 8. *Cont.*

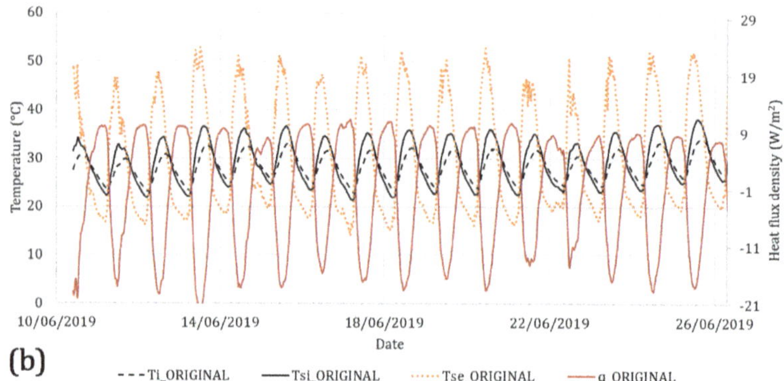

Figure 8. Heat fluxes, indoor temperatures and internal surface temperatures registered during summer for the green roof (**a**) and the original one (**b**).

On the contrary, for the original roof, an average indoor temperature of 28.93 °C and an average internal surface temperature of 29.87 °C can be noticed. Regarding the heat fluxes, in the green roof a mean value of 2.19 W/m^2 was registered. On the contrary, in the original roof, a negative mean value of −1.86 W/m^2 highlighted the incoming direction of the heat flows. Analyzing the external surfaces' temperatures, the original roof reached 50 °C, while the external temperatures of the roof-lawn system reached values below 30 °C. As mentioned before, the influence of the thermophysical characteristics of the materials used played a fundamental role. The solar radiation was absorbed by the original roof covering tiles due to their high absorptance coefficient. This did not happen for the green roof, characterized by evapotranspiration phenomena.

As previously mentioned, the dynamic performance of the roofs can be evaluated by means of the heat waves' phase shift. PS and DF parameters were obtained analyzing the summer months, when the solar radiation provides the highest influence. The average PS of the green roof was equal to 6 h and 50 min, much higher than the original roof, with a PS of 3 h and 30 min. In terms of decrement factor, the green roof showed a DF equal to 0.19, while the original roof was characterized by a DF equal to 0.37. It is essential to have a thermal wave phase shift of at least 8 h, or of no less than 10 h in areas characterized by a hot summer. The phase shift value, often neglected during the design phase, is surely critical for determining summer thermal comfort, with effects in terms of energy savings. In summer, the heat stored by the envelope is gradually released inside the rooms with a time delay that attenuates and postpones the heat peak, thus reducing the cooling energy needs. Here, the roof-lawn system is simply placed on a reinforced concrete slab which was not optimized to obtain the best performance of the roof. For this reason, future developments will concern the optimization of the structural part of the roof, designing a stratigraphy able to work with the overlying roof-lawn.

3.2. Equivalent Thermophysical Properties

On-site measurement data were used for generating a model through Comsol software. Thus, different equivalent thermophysical properties were iteratively tested and the internal surface temperatures were simulated. The search for equivalent thermophysical properties aimed at finding the best reproduction of the green roof behavior, trying to get the best match between internal measured and simulated surface temperatures. The best matching is shown in Figure 9a, where the comparison between measured (green line) and simulated (black dotted line) internal surface temperatures is reported.

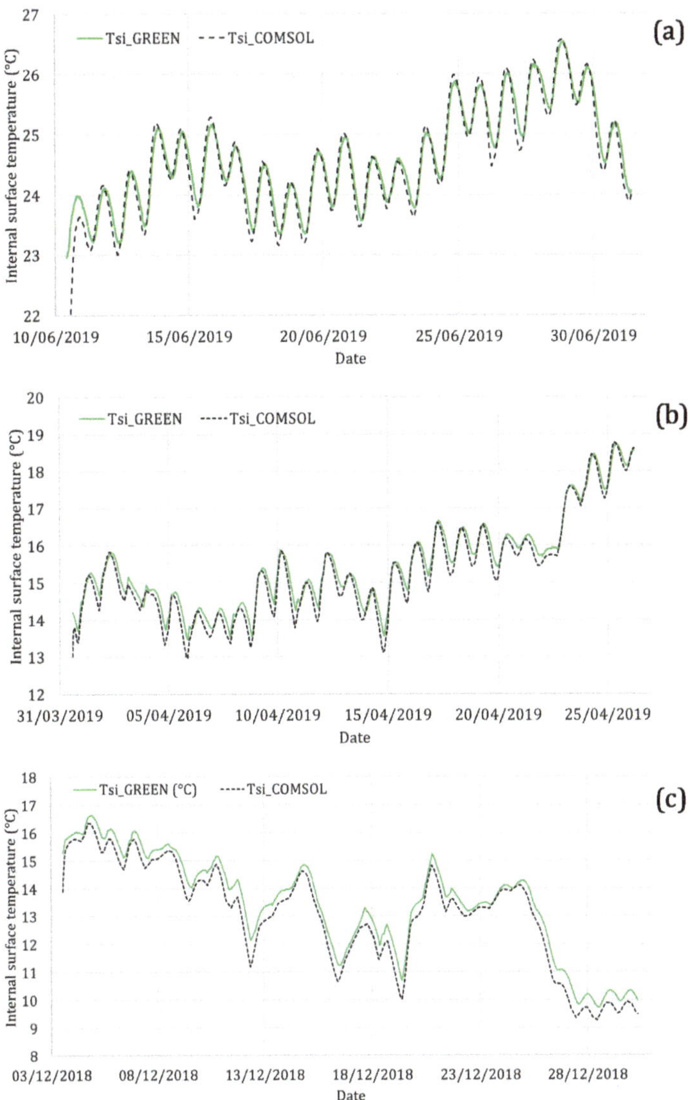

Figure 9. Green roof equivalent model. Comparison between measured and simulated internal surface temperature: summer (**a**); middle season (**b**); winter (**c**).

During summer, the overlap between experimental and simulated data was found, setting the following equivalent parameters: thermal conductivity equal to 0.4 W/(mK), specific heat capacity equal to 840 J/kgK and mass density equal to 1100 kg/m^3. The equivalent thermophysical properties mentioned before were also tested during middle season (April) and winter (December), when the external climatic conditions are different (see Figure 9b,c). The EF coefficients were calculated in all the mentioned seasons: during summer, EF = 0.96 was obtained; during the middle season, EF = 0.96 was computed; and, finally, during winter, EF = 0.93 was found.

The EF values are all higher than 0.9, satisfying the desired condition reported in the methodology section (see the flow-chart in Figure 4). The equivalent thermophysical properties can therefore be used in building energy simulation tools.

3.3. Building Energy Simulations

The equivalent thermophysical properties defined in the previous section were used as inputs in the building energy software, in order to simulate the influence of the green roof on the annual energy needs. As mentioned in the methodology section, a simple detached building was created by means of TRNSYS software.

Comparing the effects related to the installation of the green roof respect to the original one, Figure 10 shows the difference between the annual heating and cooling energy needs. It is possible to observe a percentage difference equal to −21.14% for heating and −34.70% for cooling.

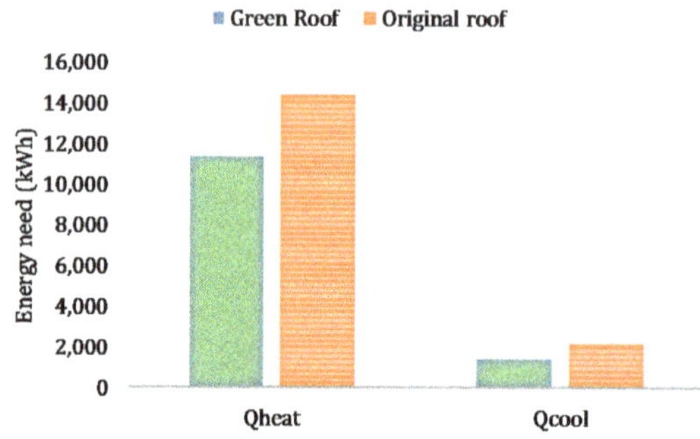

Figure 10. Comparison between annual heating and cooling energy needs.

In addition, the energy simulations were performed, taking into account different climatic conditions. Following the climatic classification reported in [28], Table 2 lists the heating and cooling energy needs obtained using four different weather data: Rome, Manaus, Abu Dhabi and Moscow were considered as references for mild temperate, tropical, dry and snowy conditions, respectively. Comparing the green and the original roofs, the values reported in Table 2 always obtain negative percentage variations, highlighting that the green roof could be applied under different climatic conditions, showing positive effects during winter and summer.

Table 2. Comparison among heating and cooling energy needs in different climatic conditions.

	Rome		Manaus		Abu Dhabi		Moscow	
	Heating	Cooling	Qheat	Qcool	Qheat	Qcool	Qheat	Qcool
Green roof	11,388 kWh	1439 kWh	0 kWh	4798 kWh	379 kWh	10,597 kWh	34,529 kWh	42 kWh
Original roof	14,442 kWh	2205 kWh	0 kWh	7168 kWh	654 kWh	14,197 kWh	42,807 kWh	113 kWh
Variation	−21.14%	−34.70%	-	−33.06%	−42.03%	−25.36%	−19.34%	−62.90%

Starting from the obtained results, it is possible to affirm that the green roof has a good insulating effect, reducing the energy needs of the building in cold seasons and keeping it cooler under warm climatic conditions (due to its inertial behavior). It is worth mentioning that a green roof can also protect the roof's materials from temperature fluctuations, ensuring the transpiration of the layers.

4. Conclusions

A green roof, installed on an actual building, was examined through on-site measurements. Several measuring instruments were applied for monitoring heat transfers phenomena across the green roof, making a comparison with a nearby conventional roof. Due to difficulties modelling the green part of the roof, an equivalent model was generated for obtaining equivalent thermophysical properties to be used in building energy simulation tools.

The outcomes of this study conclude that:

- The comparison between the original and the green roof reveals that the roof-lawn system has a more stable thermal behavior, during both summer and winter seasons;
- Making a comparison between the green and the original roofs U-values, a percentage difference of about −55% was highlighted, demonstrating a significant insulating effect of the green roof;
- The roof-lawn system significantly increases the inertial behavior of the roof, generating higher thermal comfort in the indoor environment for occupants;
- The equivalent thermophysical properties were found and verified during summer, middle season and winter, thus demonstrating the effectiveness and the reliability of the assessed values;
- A significant reduction in the energy needs of the building was achieved: when simulating the roof-lawn system compared to the original roof, percentage differences of −21.14% and −34.70% were obtained for heating and cooling, respectively.

Green roofs have a thermal insulation function, known since ancient times. It is therefore true that this kind of systems involve higher initial costs, which can be amortized quickly [7]. Green roofs are also a natural barrier against noise pollution. The green roof can absorb external noise by reducing the reflection of sound. In addition, greenery produces oxygen and captures CO_2 and polluting agents, representing a natural countermeasure against air pollution. Among passive solutions, green roofs represent a sustainable answer, reducing the collateral effect of the Urban Heat Island phenomenon.

Future developments will concern the optimization of the structural part of the roof, and designing a stratigraphy able to better work with the overlying roof-lawn.

Author Contributions: Data curation, C.G. and L.E.; Formal analysis, C.G. and L.E.; Methodology, C.G., L.E., F.A. and R.D.L.V.; Supervision, F.A. and R.D.L.V.; Writing—original draft, C.G. and L.E.; Writing—review & editing, F.A. and R.D.L.V. All authors have read and agreed to the published version of the manuscript.

Funding: This research received no external funding.

Conflicts of Interest: The authors declare no conflict of interest.

Nomenclature

C	Thermal conductance [W/(m²K)]
h_{int}	Internal heat transfer coefficient [W/(m²K)]
q	Heat flux density [W/m²]
T	Temperature at the boundary of the geometry [K, °C]
T_e	Outdoor air temperature [K, °C]
T_{env}	Temperature outside the simulated domain [K, °C]
T_i	Indoor air temperature [K, °C]
T_{se}	External surface temperature [K, °C]
T_{si}	Internal surface temperature [K, °C]
U	Thermal transmittance [W/(m²K)]
T	Time [h, min]
MAX	Maximum value
MIN	Minimum value
PS	Phase shift [h]
DF	Decrement factor [-]

References

1. Guattari, C.; Evangelisti, L.; Balaras, C. On the assessment of urban heat island phenomenon and its effects on building energy performance: A case study of Rome (Italy). *Energy Build.* **2018**, *158*, 605–615. [CrossRef]

2. Mohajerani, A.; Bakaric, J.; Jeffrey-Bailey, T. The urban heat island effect, its causes, and mitigation, with reference to the thermal properties of asphalt concrete. *J. Environ. Manag.* **2017**, *197*, 522–538. [CrossRef] [PubMed]

3. Rizwan, A.M.; Dennis, L.Y.C.; Liu, C. A review on the generation, determination and mitigation of Urban Heat Island. *J. Environ. Sci.* **2008**, *20*, 120–128. [CrossRef]

4. Castiglia Feitosa, R.; Wilkinson, S.J. Attenuating heat stress through green roof and green wall retrofit. *Energy Build.* **2018**, *140*, 11–22. [CrossRef]

5. Jim, C.Y. Green roof evolution through exemplars: Germinal prototypes to modern variants. *Sustain. Cities Soc.* **2017**, *35*, 69–82. [CrossRef]

6. Besir, A.; Cuce, E. Green roofs and facades: A comprehensive review. *Renew. Sustain. Energy Rev.* **2018**, *82*, 915–939. [CrossRef]

7. Shafique, M.; Kim, R.; Rafiq, M. Green roof benefits, opportunities and challenges—A review. *Renew. Sustain. Energy Rev.* **2018**, *90*, 757–773. [CrossRef]

8. Solcerova, A.; van de Ven, F.; Wang, M.; Rijsdijk, M.; van de Giesen, N. Do green roofs cool the air? *Build. Environ.* **2017**, *111*, 249–255. [CrossRef]

9. Imran, H.M.; Kala, J.; Ng, A.W.M.; Muthukumaran, S. Effectiveness of green and cool roofs in mitigating urban heat island effects during a heatwave event in the city of Melbourne in southeast Australia. *J. Clean. Prod.* **2018**, *197*, 393–405. [CrossRef]

10. Yang, J.; Kumar, D.I.M.; Pyrgou, A.; Chong, M.; Santamouris, D.; Kolokotsa, S.E. Lee, Green and cool roofs' urban heat island mitigation potential in tropical climate. *Sol. Energy* **2018**, *173*, 597–609. [CrossRef]

11. Zhang, Q.; Liping, M.; Wang, X.; Liu, D.; Zhou, B.; Zhu, L.; Sun, J.; Liu, J. The capacity of greening roof to reduce stormwater runoff and pollution. *Landsc. Urban Plan.* **2015**, *144*, 142–150. [CrossRef]

12. Wang, H.; Qin, J.; Hu, Y. Are green roofs a source or sink of runoff pollutants? *Ecol. Eng.* **2017**, *107*, 65–70. [CrossRef]

13. Teotonio, I.; Matos Silva, C.; Oliveira Cruz, C. Eco-solutions for urban environments regeneration: The economic value of green roofs. *J. Clean. Prod.* **2018**, *199*, 121–135. [CrossRef]

14. Huang, Y.Y.; Chen, C.T.; Liu, W.T. Thermal performance of extensive green roofs in a subtropical metropolitan area. *Energy Build.* **2018**, *159*, 39–53. [CrossRef]

15. Khabaz, A. Construction and design requirements of green buildings' roofs in Saudi Arabia depending on thermal conductivity principle. *Constr. Build. Mater.* **2018**, *186*, 1119–1131. [CrossRef]

16. Ziogou, I.; Michopolous, A.; Voulgari, V.; Zachariadis, T. Implementation of green roof technology in residential buildings and neighborhoods of Cyprus. *Sustain. Cities Soc.* **2018**, *40*, 233–243. [CrossRef]

17. Asdrubali, F.; Evangelisti, L.; Guattari, C.; Marzi, A.; Roncone, M. Monitoraggio e simulazione dinamica di un edificio pilota dotato di tette verd. *Aicarr. J.* **2019**, *59*, 40–44. [CrossRef]

18. Liu, K.; Baskaran, B. Thermal performance of green roofs through field evaluation. In Proceedings of the First North American Green Roof Infrastructure Conference, Awards and Trade Show, Chicago, IL, USA, 29–30 May 2003.

19. Liu, K.; Baskaran, B. Thermal Performance of extensive green roofs in cold climate. In Proceedings of the 2005 World Sustainable Building Conference, Tokyo, Japan, 27–29 September 2005.

20. Bindi Prato Pronto. Available online: http://www.pratopronto.it/index.php?option=com_content&view=article&id=51&Itemid=57 (accessed on 3 March 2020).

21. ISO 9869-1. *Thermal Insulation: Building Elements—In-Situ Measurement of Thermal Resistance and Thermal Transmittance. Part 1: Heat Flow Meter Method*; ISO: Geneva, Switzerland, 2015.

22. Kontoleon, K.J.; Bikas, D.K. The effect of south wall's outdoor absorption coefficient on time lag, decrement factor and temperature variations. *Energy Build.* **2007**, *39*, 1011–1018. [CrossRef]

23. COMSOL Multiphysics, version 5.2. Available online: www.comsol.com (accessed on 9 January 2020).

24. Guattari, C.; Evangelisti, L.; Gori, P.; Asdrubali, F. Influence of internal heat sources on thermal resistance evaluation through the heat flow meter method. *Energy Build.* **2018**, *135*, 187–200. [CrossRef]

25. Evangelisti, L.; Guattari, C.; Gori, P.; Asdrubali, F. Assessment of equivalent thermal properties of multilayer building walls coupling simulations and experimental measurements. *Build. Environ.* **2018**, *127*, 77–85. [CrossRef]

26. Evangelisti, L.; Guattari, C.; Gori, P.; de Lieto Vollaro, R.; Asdrubali, F. Experimental investigation of the influence of convective and radiative heat transfers on thermal transmittance measurements. *Int. Commun. Heat Mass* **2016**, *78*, 214–223. [CrossRef]
27. Levinson, R. Cool Roofs, Cool Cities, Cool Planet. In Proceedings of the China's National Development and Reform Commission Delegation to LBNL, Berkeley, CA, USA, 22 April 2010.
28. Evangelisti, L.; Asdrubali, G.F. On the sky temperature models and their influence on buildings energy performance: A critical review. *Energy Build.* **2019**, *183*, 607–625. [CrossRef]

Article

3D Thermal Imaging System with Decoupled Acquisition for Industrial and Cultural Heritage Applications

Ivo Campione [1,*], Francesca Lucchi [1], Nicola Santopuoli [2] and Leonardo Seccia [3]

[1] Department of Industrial Engineering, University of Bologna, Via Fontanelle 40, 47121 Forlì (FC), Italy; f.lucchi@unibo.it

[2] Department of History, Representation and Restoration of Architecture (Italian acronym DSDRA), Sapienza University of Rome, Piazza Borghese 9, 00186 Rome, Italy; nicola.santopuoli@uniroma1.it

[3] Department of Mathematics and CAILab (Archeo-Engineering Laboratory), University of Bologna, Via Fontanelle 40, 47121 Forlì (FC), Italy; leonardo.seccia@unibo.it

* Correspondence: ivo.campione@unibo.it

Received: 18 December 2019; Accepted: 22 January 2020; Published: 23 January 2020

Abstract: Three-dimensional thermography is a recent technique—with various fields of application—that consists of combining thermography with 3D spatial data in order to obtain 3D thermograms, high information objects that allow one to overcome some limitations of 2D thermograms, to enhance the thermal monitoring and the detection of abnormalities. In this paper we present an integration methodology that can be applied to merge data acquired from a generic thermal camera and a generic laser scanner, and has the peculiarity of keeping the two devices completely decoupled and independent, so that thermal and geometrical data can be acquired at different times and no rigid link is needed between the two devices. In this way, the stand-alone capability of each device is not affected, and the data fusion is applied only when necessary. In the second part, the real effectiveness of our approach is tested on a 3D-printed object properly designed. Furthermore, one example of an application of our methodology in the cultural heritage field is presented, with an eye to preservation and restoration: the integration is applied to a marble statue called Madonna with the Child, a fine work of the Florentine sculptor Agostino di Duccio (1418–1481). The results suggest that the method can be successfully applicable to a large set of scenarios. However, additional tests are needed to improve the robustness.

Keywords: 3D thermography; thermal imaging; laser scanning; integration methodology; extrinsic calibration; decoupled acquisition; heritage conservation

---

1. Introduction

All bodies at temperature above absolute zero (0 K) emit electromagnetic radiation. If the temperature of a body has the same magnitude as the ambient temperature, then the emission is mainly relegated to the infrared (IR) range of the spectrum and can be sensed and displayed by a thermal camera as a false-color image, called a thermogram. In addition, by knowing a series of parameters such as the inspected object emissivity and the apparent reflected temperature (i.e., the ambient temperature), an approximated temperature map of the object can be computed in output. Infrared thermography is a non-invasive (non-contact and non-destructive) imaging method, which makes it a widely applicable technique. For example, it has a vast range of applications in research and industry, building and infrastructure, electrical installation inspection, microsystems engineering, but also in biology, medicine, life sciences and cultural heritage [1,2].

Like thermal imaging, 3D reconstruction techniques are nowadays widespread in many different fields and are commonly used to acquire the object geometry and to provide easy 3D documentation.

They are a powerful tool to improve the identification, monitoring, conservation and restoration of objects and structures [3].

The utility of the integration is largely due to the fact that the ability to combine both temperature and geometric data together can lead to several advantages: it enhances and speeds up the interpretation of the results; it offers the possibility to select the region of interest by taking into account the geometry; it allows the easy segmentation of the 2D data from the background. One of the most important advantages, however, is that it allows one to overcome a significant limitation of 2D thermograms, namely the systematic error in the measured temperature due to the dependence of the emissivity on the viewing angle [4–6].

Several works in the literature have shown the strong potential of 3D thermal mapping (commonly known in the literature as 3D thermography). The thermal data are obtained with a thermal camera, whereas the way in which 3D data are acquired varies: a concise overview can be found in the work of G. Chernov et al. [4]. For example, in the medical field some applications of 3D thermography are the work of X. Ju et al. [7] and the one of K. Skala et al. [8]. In [7], the process of 3D capture relied upon stereo photogrammetry, whereas in [8] the system consisted of a high-resolution offline 3D laser scanner and a real-time low-resolution 3D scanner, both paired together with a thermal imaging camera, for human body 3D thermal model comparison and analysis. In [9], S. Vidas and P. Moghadam presented "HeatWave", a handheld, low-cost 3D thermography system, which allows non-experts to generate detailed 3D surface temperature models for energy auditing. The core technology of this device is obtained by combining a thermal camera and an RGB-D sensor (depth sensing device coupled with an RGB camera). In several other latest applications, such as [10–12], the spatial data were obtained by using a depth camera such as the Microsoft Kinect, which has become one of the top choices for 3D thermography, because of its large versatility and the capability to be exploited to perform real-time integrations. In [13], the integration was carried out on the data acquired by two smartphones arranged in a stereo configuration and a thermal camera. In [14], a fully automatic system that generated 3D thermal models of indoor environments was presented; it consisted of a mobile platform equipped with a 3D laser scanner, an RGB camera and a thermal camera.

In the cultural heritage field, spatial and multispectral data have usually been fused together for documentation reasons, historical studies, restoration plans and visualization purposes; several examples can be found in [15–19].

One advantage of 3D thermal models is that, for each 3D point, one can compute the so-called viewing angle (i.e., the angle between the surface normal vector in that point and the vector joining the point and the optical center). This information can be used to correct the error in the temperature caused by the dependence of the emissivity on the viewing angle. Indeed, for a given material, the emissivity is usually not constant, but depends on several factors, such as the surface condition, the wavelength, the temperature, the presence of concavities and the viewing angle (a viewing angle-dependent emissivity is often called "directional emissivity"). A detailed explanation of how these factors affect the emissivity can be found in [1] (pp. 35–45). Whereas the role of many of these factors can be in general considered negligible, the dependence on the viewing angle is normally relevant, and can bias the results, as outlined, for example, in [20] and [5]. Therefore, by knowing both the directional emissivity and viewing angle, it is possible to correct the temperature accordingly. Examples can be found in [4,6] and moreover in [21], where the internal reflections due to concave surfaces of a complex test setup were also taken into account.

It is worth noticing that in these publications the different sensors were rigidly linked together (mainly for calibration purposes and real-time data integration), and the trend was to strengthen this physical union (until obtaining, in the final form, a unique device such as "HeatWave" [9]). However, in some cases, it can be more convenient to keep the two devices decoupled and independent. This is especially true in outdoor surveys, where there is often the need to perform the thermographic analysis at a specific day-time or night-time and/or weather conditions, which requires high versatility (e.g., for the assessment of the damages and energy efficiency of the building envelope [18]). Laser scanners, on the other hand, can be bulky and heavy; their handling and the regulation of their position and orientation (usually are mounted on a tripod) may be time consuming and requires caution. Fixing a thermal camera to this type of scanner would make them even more difficult to regulate, and the easiness of handling of the thermal camera would be compromised. Furthermore, the two devices may have very different optics, which make the optimal distance of acquisition distinct. Conversely, with a decoupled acquisition, the integration can be applied in a flexible way, namely only when it is useful, based on previously recorded data, and does not affect the stand-alone capability of each device. In the literature, works based on this approach are uncommon. One exception is the work of A. G. Krefer et al. [22], which consisted of a method for generating 3D thermal models with decoupled acquisition, which relies on structure from motion and particle swarm optimization. Our paper focuses as well on a decoupled type of integration, but it differs from [22] in several aspects, such as the calibration method, the data fusion technique and the management of the superimposition of multiple thermograms.

This paper is organized as follows: the system architecture is outlined in Sections 2–4 explain the geometrical calibration and data fusion procedure adopted, respectively, and Section 5 presents the results. In the first experimental case of Section 5, the effectiveness of our approach is tested on a 3D-printed object properly designed. The second example of application belongs to the cultural heritage field. In the last few years, we have carried out a many-sided research project aimed at preserving and restoring the ancient sanctuary of Santa Maria delle Grazie, built toward the middle of the 15th century, in the place of Fornò near the city of Forlì (Italy). In particular, we have applied thermal imaging and laser scanning both to the building at large and to the ornamental elements. One example, presented in this paper, is the application to a marble statue called Madonna with the Child, an admirable work of the Florentine sculptor Agostino di Duccio (1418–1481), made up of four superimposed blocks. Up to the year 2000, this sculpture was in a niche on the entrance arch to the prothyrum of the sanctuary. Afterwards, since it showed clear signs of deterioration, especially due to rainwater and air pollution, it was carefully restored and then moved permanently to a great hall in the Bishop's palace in Forlì, where our surveys were performed.

2. System Architecture

In this section, the system architecture is presented, with the scope to define precisely the system components and the followed workflow.

2.1. System Components

The experimental set-up consisted of a thermal camera Testo 882 and a triangulation laser scanner Konica Minolta Vivid 9i, both shown in Figure 1. The Testo 882 has an FPA detector type with 320 × 240 pixel (but image output up to 640 × 480 thanks to the super-resolution feature), a FOV of 32° × 23° and a range of detected temperature switchable between −20 °C and +100 °C and 0 °C and +350 °C (accuracy ±2% of reading for both). The Konica Minolta specifications are 307,200 pixels, three interchangeable lens of focal length 25 mm, 14 mm and 8 mm, and a weight of approximately 15 kg.

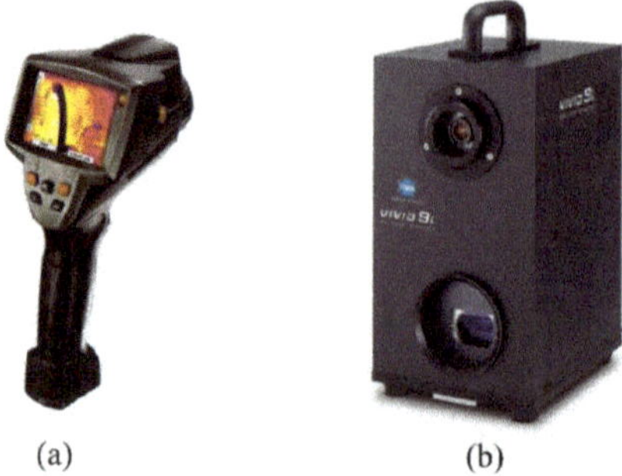

(a)　　　　　　　(b)

Figure 1. (**a**) Thermal camera Testo 882. (**b**) Laser scanner Konica Minolta Vivid 9i.

2.2. Integration Process Workflow

The workflow includes the processes for combining spatial data, acquired by the laser scanner in the form of a point cloud (spatial coordinates and normal vectors in each point) and the two-dimensional temperature map provided by the thermal camera, available as a temperature matrix. Before proceeding to the actual integration, it was necessary to have a post-processed point cloud, i.e., the registration of several range maps (by the well-known ICP algorithm) and all the post-processing operations have to be done previously.

There are three main processes characterizing the workflow (see Figure 2): acquisition, geometrical calibration (subdivided into intrinsic and extrinsic calibration) and data fusion. The emphasis of this work was both on the extrinsic calibration and on the data fusion process, which were carried out by Matlab programming (Supplementary Materials).

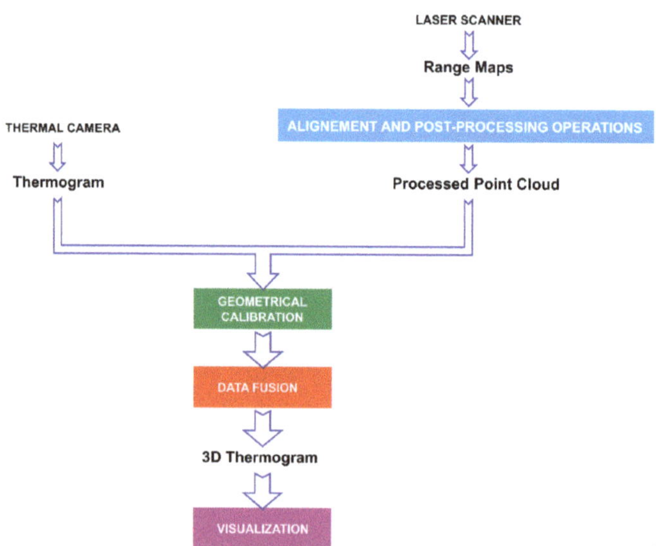

Figure 2. High-level workflow of the integration process.

3. Geometrical Calibration

When a thermal camera is involved, the term "calibration" could refer to two different types of calibration—the geometrical calibration and the radiometric calibration. The latter (which was not the object of this work) consists of a procedure that models the relationship between the digital output of the camera and the incident radiation. Hereinafter, if the term "calibration" is used, "geometrical calibration" is intended.

The geometrical calibration is a fundamental process used to compute intrinsic and extrinsic parameters of the system components.

Intrinsic parameters model the optic imaging process and are divided into two groups—four parameters defining the camera matrix, based on the pinhole camera model, and other parameters defining radial and tangential camera distortions caused by the physical properties of lenses [23].

There are six extrinsic parameters which describe the relative pose of the sensors, i.e., the transformation from world coordinates to camera coordinates. In the vast majority of methods, the intrinsic parameters are computed first, and then they are exploited to determine extrinsic ones.

For the determination of both intrinsic and extrinsic parameters, the best-known techniques involve the use of a target with a chessboard pattern, a method first presented by Z. Zang in [24]. Note that, in the case of thermal cameras, the pattern has to be detected in the IR spectrum. Several solutions have been presented for this purpose, involving, for example, a heated chessboard pattern or marker detectable in IR; a more detailed state of art can be found in [10]. In the last few decades, however, some work has been done in order to develop the so-called automatic or targetless extrinsic calibration methods, that exploit natural scenes' features (see [11,25–27]).

Concerning the intrinsic calibration, we adopted the method developed by S. Vidas et al. in [28], which relies on a mask-based approach and does not require specialized equipment.

Regarding the extrinsic calibration, since our aim was to keep the two devices unlinked, the classical approach (consisting of acquiring several frames of the target in a fixed configuration of the two sensors) did not appear very suitable. For cases in which the identification of some homologous point is easily feasible, the extrinsic parameters are computed by means of a manual selection routine of homologous points, and then by exploiting the Matlab function "estimateWordCameraPose", which solves the perspective-n-point (PnP) [29]. However, this method might fail or achieve a very low accuracy in scenarios in which homologous points cannot be clearly identified. To address this limitation and to be able to consider more various scenarios, we developed an alternative method exploiting the detection of the object silhouette in the thermogram. In the literature, some silhouette-based calibration methods can be found, for instance in [30] and [31]. In [30], the pose of a known object is estimated through a hierarchical silhouette matching and unsupervised clustering, whereas in [31] the calibration of a stereo camera system was carried out by defining and minimizing a function that computes the distance between viewing cones, which are known from the silhouettes. The developed method has some similarities with the one presented in [11], in which the extrinsic parameters were obtained by minimizing an objective function that measures the alignment between edges extracted from RGB-D images (obtained by a depth camera) and the ones extracted from thermograms. Like that method, ours is suitable for thermograms in which the object contour can be easily identified, and so the object can be extracted from the background, a fairly common situation when dealing with thermograms. Differently from [11], however, the object contour detection is only needed for the thermogram, and the quantity evaluated is not the alignment between thermal and depth edges, but rather the "degree of filling" of the projected 3D points inside the thermal edge, which translates to a different definition of the objective function to minimize. Another difference is that this method takes as its input the full 3D representation of the object, and not the single view representation (often referred to as 2.5D [32]). This method can deal efficaciously with initial parameter values significantly different from the ones searched for, which makes it particularly suitable for a decoupled type of integration.

The extrinsic calibration method here developed can be divided into the following principal parts:

(1)   extraction of the object contour from the thermogram;

(2)   creation of a matrix $\overline{\overline{M}}_1$ (with dimension equal to the resolution of the thermogram) obtained in this way: starting from the contour extracted in 1), a series of internal and external subsequent layers are created, and the same numerical value is associated to each pixel of each layer, following a specific function f_l defined later in this section;

(3)   projection of all the 3D points on the image plane with an initial set of extrinsic parameters, and the creation of a second matrix $\overline{\overline{M}}_2$ composed of binary values, 1 if on that pixel at least one 3D point has been projected, 0 otherwise. Furthermore, in order to have a more uniform projection, the pixels sufficiently close to each pixel with values 1 are set to 1 if 0 (to avoid non uniform situations, especially for sparse point clouds);

(4)   minimization of the objective function of Equation (1), with a global minimization technique:

$$f_{obj} = f_{obj}\left(\overline{\overline{R}}, \overline{t}\right) = -sum\left\{\overline{\overline{M}}_1 \circ \overline{\overline{M}}_2\left(\overline{\overline{R}}, \overline{t}\right)\right\} \tag{1}$$

where $\circ$ is the Hadamard product, *sum* is the operator that sums all the elements in a matrix, $\overline{\overline{R}}$ is the rotation matrix (computed as a function of the three Euler angles) and $\overline{t}$ is the translation vector.

The Euler angles and the components of the vector $\overline{t}$ that make the function f_{obj} minimum are the searched values, namely the extrinsic parameters. It is to be outlined that the matrix $\overline{\overline{M}}_1$, once computed, remains fixed in the optimization process described in Equation (1) and so only one contour extraction operation is required.

The point (1) can be done either in an automatic way (exploiting some well-known edge detection or background extraction algorithm) or manually. The background extraction also permits us to avoid the possible gross error of assigning to the 3D points, if the corresponding pixels are very near to the edges, the background temperature.

For the point (2), the function f_l has to be properly chosen, so to assure that the objective function minimum corresponds to the desired situation, namely that the 3D projected points completely fill up the zone inside the contour without overflowing outside the contour.

The function f_l utilized is of Gaussian type:

$f_l(x) = c_1 e^{-c_2 x^2}$ if $x \geq 0$ (layer x inside the contour or the contour itself if $x = 0$)

$f_l(x) = c_1 e^{-c_2 x^2} - c_3$ if $x < 0$ (layer x outside the contour)

($x \in Z$, c_1, c_2, c_3 positive constants).

An example of the application of the function f_l is shown in Figure 3.

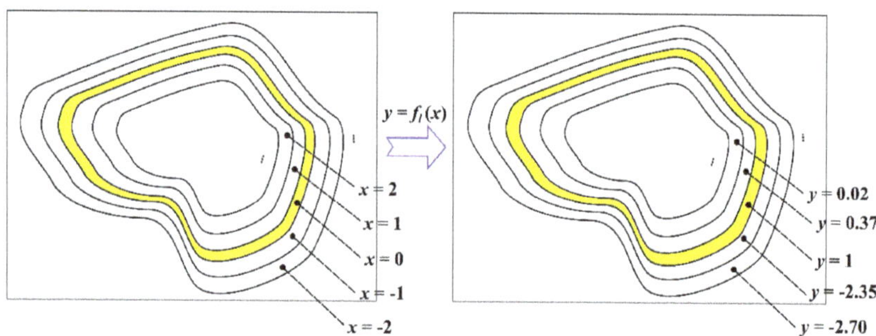

Figure 3. Example of the application of the function f_l on the layers map (with $c_1 = c_2 = 1$, $c_3 = e$). The object contour is outlined in yellow.

The c_3 constant is important, because it permits us to shift the curve of the external layers so that the function f_l has negative values there. This is necessary to avoid situations in which the minimum of the f_{obj} corresponds to the filling of zones incorporating the contour, but it is larger than the contour itself (in other words, to avoid that the projected points "overflow" the contour). As can be clearly seen in Figure 4, the function increases when approaching the object contour ($x = 0$) from both negative and positive values of x. This is another requirement, that the function f_l has to meet to grant the proper convergence of the function f_{obj}. Apart from these requirements, one has a certain freedom in the choice of the function f_l. In fact, the aforementioned Gaussian type function is the result of several experimental evaluations of possible functions. However, different choices of f_l can be further investigated in future works, in order to improve the speed and robustness of this calibration method.

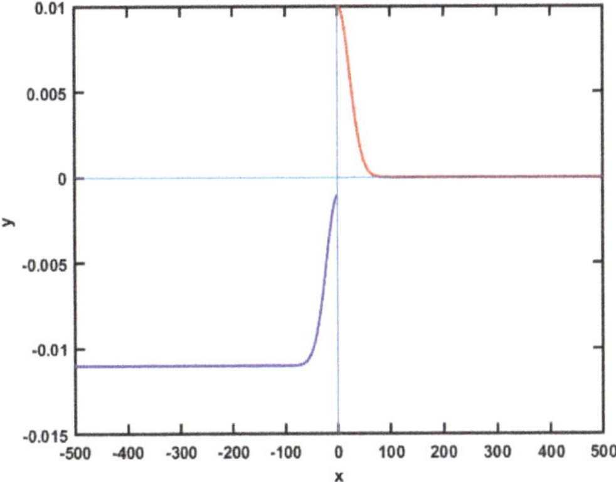

Figure 4. Graph of the function f_l (considering 500 internal and external layers) with the actual constant values utilized ($c_1 = 0.01$, $c_2 = 0.001$, $c_3 = 0.011$).

To minimize the objective function (1), with six independent variables (three Euler angles and three components of the translation vector $\bar{t}$), the Matlab optimization toolbox is exploited, in particular the function "GlobalSearch".

4. Data Fusion

In this section, the data fusion process is examined, with particular emphasis on the ray-casting technique and the multiple thermograms handling. It is assumed that all the calibration parameters have already been computed, following the steps described in the previous sections.

4.1. Ray-Casting Technique

Since the complete point cloud of the object may be utilized, it is a common situation that only a subset of all the 3D points is seen by the thermal camera when each thermogram is acquired. The process of spotting the points seen by the thermal camera is normally fulfilled by a ray-casting technique, which allows us, among other things, to handle occlusions. When working with point clouds, one common approach is to first convert the original point cloud into a different data type, such as voxels or a mesh representation. Then, to perform the visibility check, one can start from each pixel of the thermal image and move with discrete steps on a specific line, stopping when the desired voxel (or mesh triangle) is reached ([14,33]).

The approach we propose, however, does not need this kind of type conversion and works with data in form of point cloud. Starting from all the 3D points, two types of exclusion are applied in order to obtain a subset, representing all and only the 3D points seen by the camera. In the next sections, the scheme of these two exclusions will be analyzed, but first it is to be said that there is actually another exclusion, which is trivial—that is, the exclusion of all the 3D points projected outside the image plane.

4.2. First Type of Exclusion

In the first type of exclusion, the image plane is firstly divided into a grid, whose cell dimension is defined by the parameter ξ (its measurement unit is pixel, and each cell is composed of $\xi \times \xi$ pixels).

All the 3D points are projected onto the image plane, and then clustered depending on which cell of the grid they occupy. For each cluster, the points P_i, such that Equation (2) holds true, are excluded:

$$d_i > d_{min} + \chi \tag{2}$$

where d_i is the distance between the point P_i and the optical center O, d_{min} the minimum distance among the distances d_i and χ a fixed parameter (in millimeter, as for d_i and d_{min}). For higher values of ξ, higher values of χ are needed to take into account the fact that the surface region corresponding to the points not to be excluded is larger and so the difference in distances can be higher.

Theoretically, an optimal choice of the parameters ξ and χ would involve a non-trivial geometrical analysis, considering both the density in space of the 3D points (in general variable from region to region) and the curvature of the object surface with respect to the viewing angle. In practice, they can simply be adjusted by graphically evaluating the results. In our case, in order to obtain proper experimental results, they were computed according to the experimental relations Equations (3) and (4), suggesting a possible choice of the two parameters:

$$\xi = \mathrm{ceil}(Res/N) + 1\,[\mathrm{pixel}] \tag{3}$$

$$\chi = 5/(1 - \xi/20)\,[\mathrm{mm}] \tag{4}$$

where N is the number of points projected on the image plane and Res the resolution of the image. In Equation (3), "ceil" is a function that rounds a number x to the nearest integer $\geq x$. Since χ is positive, according to Equation (4) ξ must be less than 20, a condition always verified for common resolutions and points cloud densities. It has to be pointed out that, for the vast majority of object geometries, the results are stable with respect to a variation of ξ and χ of some units, and so in any case their resulting values are acceptable as long as they fall into a proper range.

4.3. Second Type of Exclusion

The second type of exclusion exploits the knowledge of the normal vectors $\hat{n}_i$ in each 3D point P_i and requires that all the points for which Equation (5) holds true are excluded:

$$(O - P_i) \cdot \hat{n}_i < 0 \tag{5}$$

where O is the optical center.

This condition derives from the fact that, if the dot product in Equation (5) is negative, the angle between the vector $(O-P_i)$ and the normal vector $\hat{n}_i$ is greater than 90 degrees, and so the point is not seen by the camera. This condition on its own is theoretically suitable for excluding all the inappropriate points in convex surfaces, but it fails when dealing with surfaces that present concavities. Conversely, even if the first exclusion appears to be able to handle a generic shape of the point cloud, it fails in some particular scenarios in which the point cloud lacks some parts, because in these cases some temperatures might be wrongly assigned to the opposite part of the point cloud. For these reasons, the two types of exclusion were combined, in accordance with the experimental results as well. Figure 5 shows two examples of applications of the two exclusions in succession with two different values of ξ, on a region of a hypothetical surface, which is represented in the points cloud by 11 points. The zones between each pair of dotted lines (converging to the optical center) represent the band corresponding to each pixel. Inside each band, the part at the right of the purple segments are excluded according to the first type of exclusion. The points in green are the points to be maintained, the points in blue are the points correctly excluded with the first type of exclusion, the points in yellow the ones correctly excluded only with the second type of exclusion and the points in red the ones not correctly excluded after applying both the exclusions, and to which consequently a wrong temperature is assigned. In Figure 5b, better results are obtained if the value of ξ is doubled, because in this way all the points belonging to a part of the object surface not seen by the camera are correctly excluded. Indeed, the value of the parameter ξ has to be chosen so that in each cell the number of reprojected points is higher than one, so that they form an actual cluster to which we can properly apply Equation (2). Finally, it is worth noticing that the order of the two exclusions is important: in fact, by inverting the exclusions order, it is easy to see that the point P_{11} is never excluded.

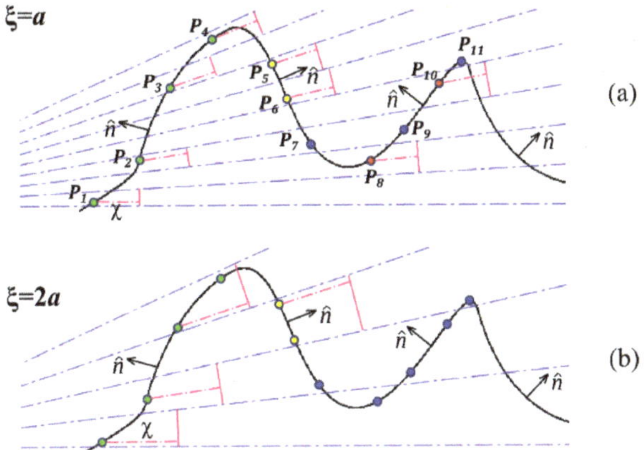

Figure 5. Application of the exclusion methods on a part of a hypothetical surface represented by eleven points, for two values of ξ. The value of ξ in (**b**) is twice the value of ξ in (**a**). Green points: to be maintained; blue points: excluded with the first exclusion; yellow points: excluded with the second exclusion; red points: not correctly excluded with the two exclusions.

4.4. Temperature Assignment

Once the subset of the 3D points seen by the thermal camera is defined, the corresponding temperature is assigned at each point, by considering in the temperature matrix the value in correspondence of the pixel in which the 3D point has been projected.

In Figure 6 the overall workflow of the data fusion process is outlined.

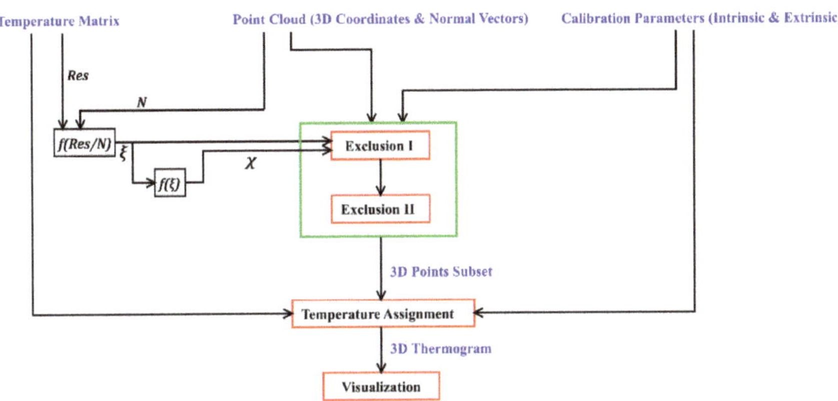

Figure 6. Overall workflow of the data fusion process.

4.5. Multiple Thermograms

In the previous sections, the case of the integration of a single thermogram on a point cloud was considered. There might be, however, the need to integrate more than one thermogram on the same point cloud, in order to have a larger set of points with an associated temperature.

Note that this procedure requires in the first place thermograms acquired in a temporal window in which no measurable changes in the thermal state occur, so as to guarantee that the integration is not carried out on discordant data. Since each thermogram can be individually integrated on the point cloud by the methodology described previously, the problem comes down to handling the overlapping zones, namely the points to which more than one temperature value is assigned. To give these points a single final temperature, the method utilized in [9] was followed. The method relies on the fact that the relative orientation between the surface of the inspected object and the camera affects the measurement accuracy, and, more specifically, the lower the viewing angle is, the higher the accuracy is. As shown in Equation (6), the temperature T_i assigned to the point P_i is computed as a weighted average of the temperatures T_{ij}:

$$T_i = \frac{\sum_{j=1}^{N_i} c_{ij} T_{ij}}{\sum_{j=1}^{N_i} c_{ij}} \tag{6}$$

where the weight is the confidence factor c_{ij}, the index i refers to the point of the points cloud and ranges from 1 to N, whereas the index j refers to the thermograms that overlap in the point i and ranges from 1 to N_i. The confidence factor c_{ij} is computed as a function of the viewing angle θ_{ij} as shown in Equation (7):

$$c_{ij} = e^{-\kappa \theta_{ij}} \tag{7}$$

where the viewing angle θ_{ij} is the angle from which the thermal camera sees the point P_i, considering the thermogram j, and can be computed as shown in Equation (8):

$$\theta_{ij} = \arccos \frac{\left(O_j - P_i\right) \cdot \hat{n}_i}{\|O_j - P_i\|} \tag{8}$$

with point O_j identifying the optical center for the thermogram j.

In this way, a greater weight is assigned to the rays with smaller viewing angles, which allows more accurate measures. More precisely, the weight decreases with an exponential law, depending on a parameter κ, that was set equal to 2 according to experimental evaluations.

4.6. Visualization

The mathematical result of the integration is a N-by-4 matrix, where each line contains the 3D coordinates of a 3D point plus the associated temperature (or, if no temperature is associated, a *NaN* value). This matrix is visualized in Matlab by assigning to the point subset with an associated temperature a particular colormap, whereas a different color is assigned to the points with no temperature associated.

Figure 7 shows an outline of the workflow in the case of the integration of multiple thermograms on the same point cloud.

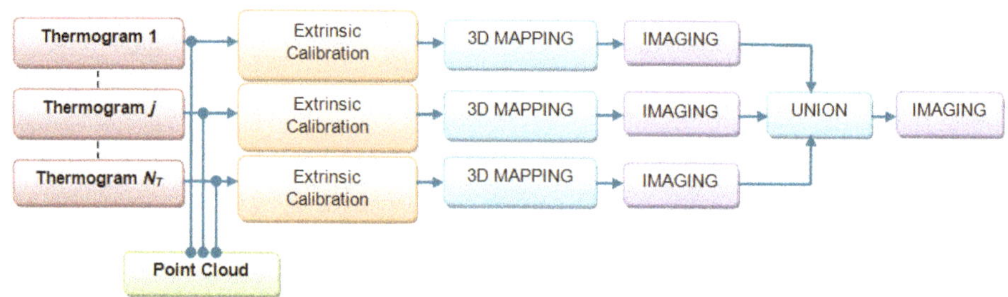

Figure 7. Workflow of the integration of multiple thermograms on the same points cloud. N_T is the total number of thermograms integrated.

5. Results

5.1. Test Object

The experiments were carried out on a particular object, designed so that the two aforementioned extrinsic calibration methods were both equally suitable and easy comparable. In addition, the particular design of the object shown in Figure 8 allowed us to evaluate the effectiveness of the ray-casting technique on a geometry with concavities. It has an internal conic hole and three different radial parts (shifted by angles of 120 degrees), characterized by many edges.

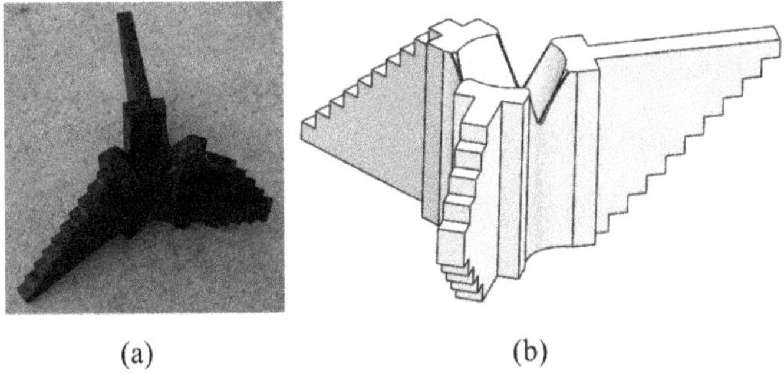

(a) (b)

Figure 8. (**a**) 3D printed object; (**b**) Its CAD model.

The object was designed with CAD, 3D-printed with acrylonitrile butadiene styrene (ABS) material and then scanned with the Konica Minolta Vivid 9i laser scanner. The point cloud could be obtained directly from the CAD model, but the object was scanned to take into account the acquisition bias and test the effectiveness of the method on laser scanner data.

Here we present a first set of simulations, carried out to assess the effectiveness of the automatic extrinsic calibration method. This was made by comparing the extrinsic parameter values obtained by means of this automatic procedure with the ones resulting from the method based on the manual selection of homologous points. Because of the specially designed geometry features of the printed object, it was possible to choose for the first method a set of about 20–30 homologous points for each analyzed thermogram, with a mean reprojection error (MRE) of about 1–2 pixels. In this way, the proper reliability of the first method parameters was assured, and so it was possible to evaluate the ones obtained with the second method. Equation (9) shows the expression for computing the MRE. The units of MRE are pixels.

$$MRE = \frac{\sum_{i=1}^{n} \|P_i - Q_i\|}{n} \tag{9}$$

In Equation (9), n is the number of homologous points selected, P_i the 2D coordinates of the *i-th* point selected on the thermogram, Q_i the 2D coordinates of the reprojection of the *i-th* homologous 3D point.

For each thermogram, the convergence from different initial positions was analyzed, within a range of ±25 degrees for the Euler angles and ±20 mm for the translation vector from the parameters obtained with the first method.

In Table 1, some of the results of the first calibration method are shown for four different thermograms, whereas Table 2 gives the differences (considering the same thermograms) between the extrinsic parameters obtained with the second method and with the first one respectively.

Table 1. Extrinsic parameters for four different thermograms, in the case of the method based on the manual selection of the homologous points (first method).

Thermogram	N° of Points	MRE	t_1 [mm]	t_2 [mm]	t_3 [mm]	α [°]	β [°]	γ [°]
A	21	1.51	−18.64	19.18	544.17	126.73	3.25	−114.01
B	26	1.51	−16.34	23.23	497.88	−73.48	−1.33	−120.38
C	21	1.61	−3.19	21.62	375.58	25.16	−0.14	−86.90
D	30	1.52	−1.11	12.48	453.16	−3.35	−1.68	−125.87

Table 2. Differences between the extrinsic parameters obtained with the automatic method (second method) and the ones obtained with the first method.

Thermogram	Δt_1 [mm]	Δt_2 [mm]	Δt_3 [mm]	$\Delta\alpha$ [°]	$\Delta\beta$ [°]	$\Delta\gamma$ [°]
A	1.13	0.01	−11.53	0.76	0.83	−0.74
B	0.78	0.96	−4.75	0.29	−0.08	1.67
C	−0.11	1.15	−7.36	0.45	−0.90	0.35
D	−0.04	0.90	−9.66	−0.07	0.41	0.43

Figure 9 shows a set of 2D points and the reprojection of 3D homologous points, once the extrinsic parameters were computed.

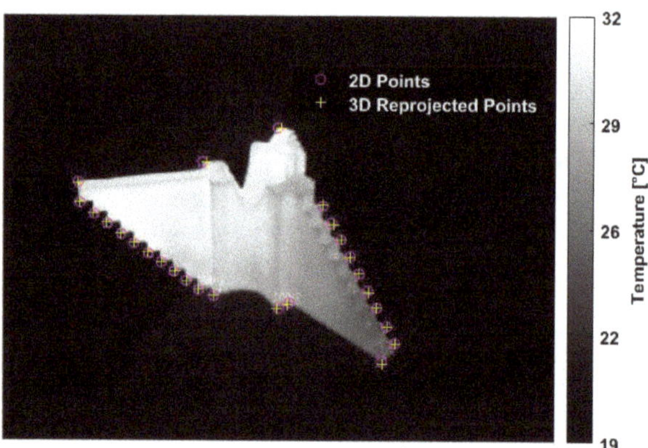

Figure 9. Grayscale thermogram with highlighted the set of 2D points (purple circles) and the reprojection of 3D homologous points (yellow cross) once the extrinsic parameters were computed, in the case of thermogram B of Table 1.

With reference to Table 2, thermograms A and B were obtained starting from a first guess of the parameters with the following differences with respect to the ones of Table 1: $\Delta t_1 = 20$ mm, $\Delta t_2 = -20$ mm, $\Delta t_3 = 20$ mm, $\Delta\alpha = -20°$, $\Delta\beta = 0°$, $\Delta\gamma = 20°$. For thermograms C and D, the initial differences were set as follows: $\Delta t_1 = -10$ mm, $\Delta t_2 = 20$ mm, $\Delta t_3 = 10$ mm, $\Delta\alpha = -20°$, $\Delta\beta = 15°$, $\Delta\gamma = 15°$.

As can be seen in Table 2, the final differences are relatively low, except for the Δt_3, which is of several millimeters. However, this can be acceptable, considering that the t_3 parameter is greater by an order of magnitude compared to the parameters t_1 and t_2 (relative error of about 2%).

In Figure 10, the initial (a) and final (b) projection in the case of automatic extrinsic calibration of the thermogram B are shown. This type of visualization allows for a qualitative consideration on the effectiveness of the method. As can be easily seen, in the case (b), the 3D projected points (in red) fill the thermogram contour (yellow line) well.

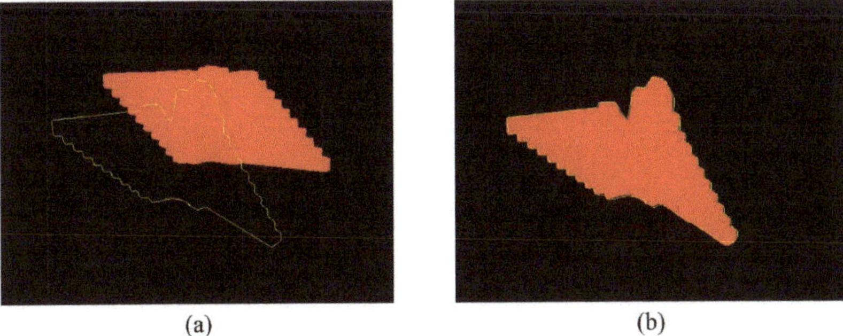

(a)   (b)

Figure 10. Case of automatic calibration of thermogram B: (**a**) Initial projection of 3D points, using a set of initial extrinsic parameters; (**b**) Projection utilizing the parameters for which the objective function of Equation (1) presents a global minimum.

Figure 11 shows a representation of the matrix $\overline{\overline{M}}_1$ (see Section 3, Figures 3 and 4) converted for visual purposes into a color image.

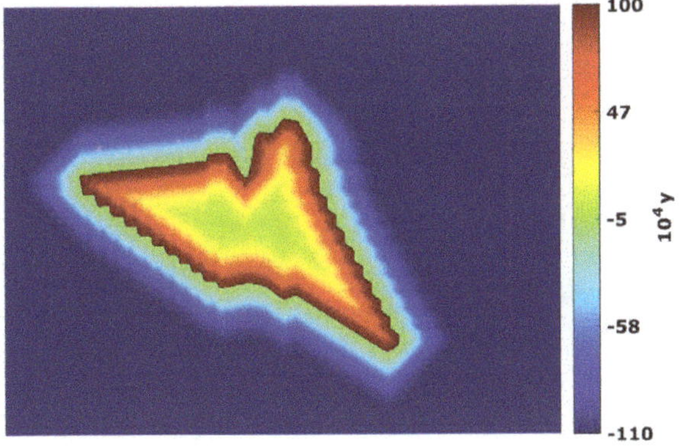

Figure 11. Visual representation of the matrix $\overline{\overline{M}}_1$ created from the yellow contour of Figure 10. In the scale, y refers to the function graphed in Figure 4, which is here multiplied by a factor of 10^4 for the sake of clarity.

The second set of experiments was aimed at evaluating the general reliability of the technique, by qualitatively assessing the final results obtained with the integration of different thermograms.

In Figure 12, the result in the case of the integration of thermogram B is shown, whereas Figure 13 shows an example of the integration of multiple thermograms (17 in total) acquired from different poses. The shots were taken while the object was being heated with a drier, fixed so that the heat flux came in contact with one side in particular (the one with higher temperature, colored yellow).

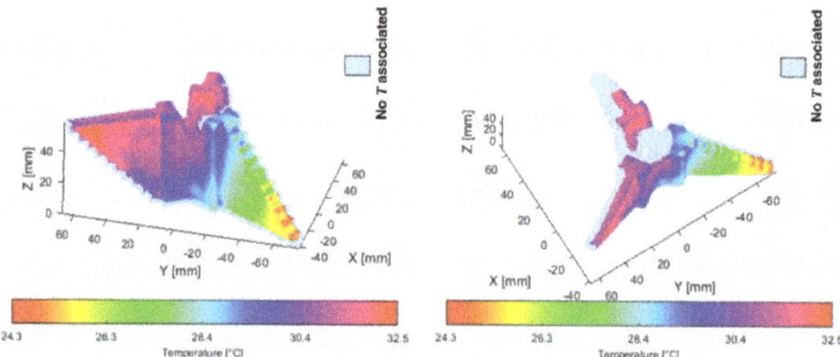

Figure 12. Two different views of the result of the integration of thermogram B on the point cloud.

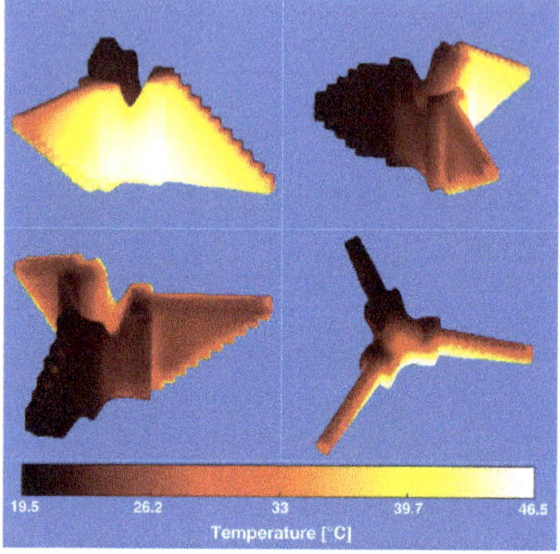

Figure 13. Four different views of the result of the integration of seventeen thermograms.

Figure 14 shows how the union of several thermograms (acquired from different poses) by the method explained in Section 4.5 can compensate for the systematic error in the temperature caused by the dependence of the emissivity on the viewing angle. In Figure 14c, the temperatures of a single thermogram and of the union of several thermograms (the temperature superimpositions vary from 4 to 8 depending on the point) are compared along a key zone, shown in Figure 14a,b, in which the normal vectors of the surface have a significant variation, which implies a significant variation on the viewing angles (referring to the case of the single thermogram, since after the union the concept of viewing angle loses meaning). In this zone, the temperature of the single thermogram is appreciably lower than the temperature of the union (with a maximum difference of roughly 1 °C). This behavior can be explained by correlating the difference ΔT of the temperatures ($\Delta T = T_{UNION} - T_{SINGLE}$) with the variation of the viewing angle, as shown in Figure 14d. The camera view during the acquisition of the single thermogram can be approximatively assumed to be the view of Figure 14a, which explains the angle of about 40 degrees for Y in the range 5–10 and 16–25 mm (the camera was tilted downwards with respect to the normal vectors in these points by about 40 degrees). For non-conductor materials, emissivity is nearly constant from 0 to 40–45 degrees, whereas at larger angles it has a significant drop [1] (pp. 39–40). This justifies the fact that the temperature in the zone of high viewing angles is underestimated in the case of the single thermogram. This error can be effectively compensated by the adopted method, as long as, for the same points, temperatures with lower viewing angles associated are available.

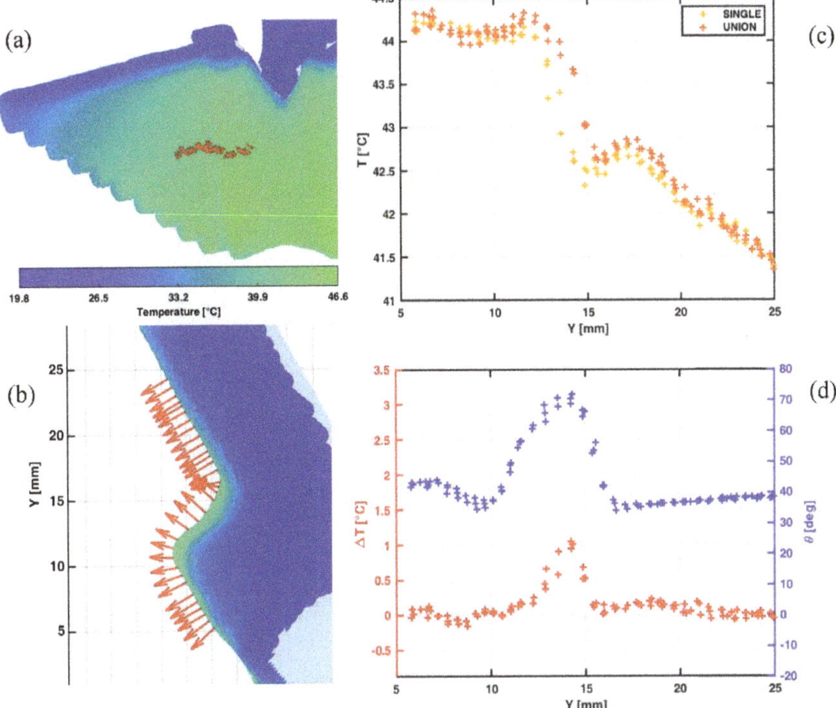

Figure 14. Effect of the temperature correction achieved by the union of different thermograms. (**a**) Zone in which the temperatures are examined, with the vectors normal to the surface in these points colored red; temperatures are shown with a "winter" colormap; (**b**) Close-up of the normal vectors, from a top view; (**c**) Comparison between the temperature of a single thermogram and the one resulting from the union of several thermograms; (**d**) Temperature difference (*T* resulting from the union minus *T* of the single thermogram) and viewing angle for each point examined.

5.2. Statue "Madonna with the Child"

After testing the method on the object previously described, in this section, an application in the cultural heritage field is considered. The integration was carried out on the statue Madonna with the Child, by the Florentine sculptor Agostino di Duccio (1418–1481), with an eye to monitoring its condition and gather additional information about its state. The point cloud of the statue was already available from previous works. Figure 15 shows the statue and the laser scanner in the Bishop's palace in Forlì. For the sake of brevity here, we present only some of the most significant results.

Figure 15. Statue Madonna with the Child, with the laser scanner in position.

In Figure 16, three thermograms of the statue acquired from different poses are shown.

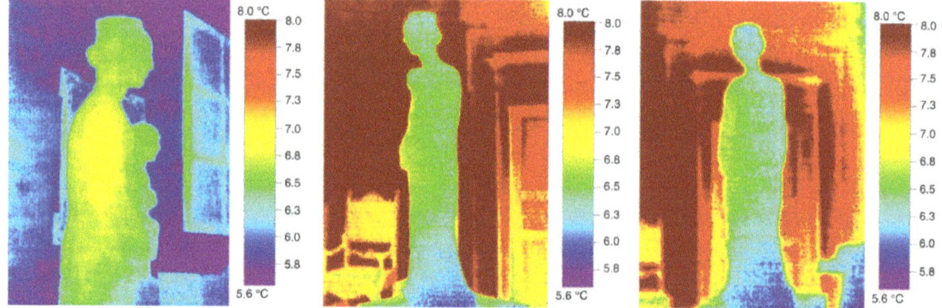

Figure 16. Three thermograms of the statue, acquired from different poses.

Figure 17 shows several outcomes of the integration. There is a little spot on the top of the head to which no temperature is assigned, and to which a specific colour not belonging to the colormap is assigned. The survey shows that there is one side of the statue that is slightly hotter, whereas the base and the head present lower temperatures. It is probable that the particular lighting system and the statue arrangement near a window with a non-insulated frame give rise to these changes, which is not favourable for optimal preservation. Further investigations are needed to better clarify the cause.

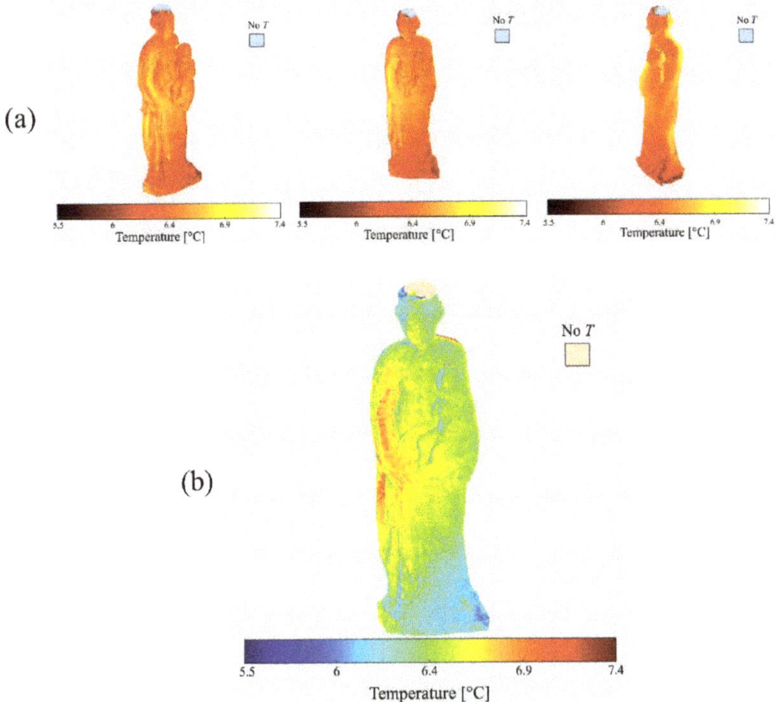

Figure 17. Results of the integration of twelve thermograms. (**a**) Three different views coloured with the colormap "hot". (**b**) Frontal view coloured with the colormap "jet".

6. Discussion

The work presented includes methodologies to address each integration step, with the final aim of achieving 3D thermography by maintaining the decoupling of the devices. Concerning the extrinsic calibration, we proposed an automatic method that exploits the object silhouette. An evaluation of the accuracy of the method is presented in Table 2. Since our automatic method does not exploit the concept of homologous points (and thus, MRE cannot be computed), the evaluation of its accuracy was made by a comparison with the extrinsic parameters (considered the ground truth) obtained by the manual selection of homologous points, executed on a test object purposely designed. A comparison can be made with the automatic calibration method based on a silhouette developed by J.T. Lussier and S. Thurn [11]. The error concerning the x and y translations (Δt_1 and Δt_2 in Table 2) in our work is about one order of magnitude lower (the max obtained is around 1 mm); concerning the z translation (Δt_3 in Table 2), our maximum error stays within 12 mm, against the over 50 mm reported in [11]. Regarding the error on the rotations ($\Delta\alpha$, $\Delta\beta$ and $\Delta\gamma$ in Table 2), our errors are higher, with a mean of 0.58 degrees against the 0.3 degrees reported in [11]. We want to point out that this comparison is made between integration procedures that suit different integration modalities. In [11], indeed, the authors carried out a real-time integration between thermograms and depth-maps (2.5D), whereas we integrate (offline) thermograms with 3D point clouds. The difference in the type of the range data integrated entails, among other things, the following fact—the accuracy of the method used in [11], as the authors said, depends on the scene coverage, that is to say the percentage of the area covered by the object of interest with respect to the background. The higher the coverage is, the higher the edge variability is, and the lower the error in the extrinsic parameters. In our case, since we did not work with an edge map but with 3D point cloud, the concept of edge variability is not similarly defined and the error did not depend on this parameter.

Unfortunately, regarding the accuracy, a direct comparison with the decoupled method presented by A. G. Krefer al. [22] is not possible because the automatic calibration procedure they used was based on the automatic matching of interest points and was evaluated by the classical MRE. Apart from the calibration method, other differences from the work [22] include the fact that they used the 3D data in the form of a polygonal mesh, whereas we keep the 3D data as a point cloud during the whole integration process. For visual purposes, if need be, the results may be converted into a coloured mesh at a later time. Concerning the handling of the points in which more temperatures are superimposed, we computed the temperature to be assigned as a weighted average, where the weight decreases in an exponential way if the viewing angle increases, which was the method exploited by S. Vidas and P. Moghadam in [9]. In [22], conversely, the weight of each point temperature was computed as a function of the position of the point inside the view frustum of the camera (the weight increases if nearer to the optical axis or to the optical centre). A flaw of this latter approach is that it did not take into account the object geometry (i.e., the normal vectors in each point) and so it was prone to fail to compensate for the variation of the emissivity at high viewing angles. Figure 14 clearly shows that at high viewing angles the temperature can be underestimated (in that specific case up to one degree). The methodology followed, which takes the object geometry into account, allows us to overcome this issue, assuming that thermograms of the interested area can be acquired from different orientations. However, if this is not possible (for instance, because of the position and of the limited mobility of the object or if the acquisition time is limited), an improvement of this method could be to apply a temperature correction to each thermogram singularly, exploiting, for instance, the correction formula proposed in [4], which relies on a theoretical model for the directional emissivity.

The whole integration methodology was first tested on a purposely designed 3D-printed object and then on a historical marble statue, and the results demonstrate the general feasibility of the approach. We are planning, however, further tests, in particular aimed at improving the robustness of the automatic extrinsic calibration method, which is affected, to a certain degree, by the object geometry (especially in terms of level of detail of the geometrical features and of the presence of symmetries). For objects which present a sufficient number of points of interest clearly identifiable both in infrared and in their 3D geometry (e.g., very sharp edges), the manual selection of the homologous points can still be a better option to compute the extrinsic parameters, and could be improved, for instance by applying to the thermogram the intensity transformation proposed in [22], which is able to highlight certain points which are otherwise not clear enough.

7. Conclusions

In this work, a decoupled acquisition method for generating 3D thermal models is proposed. The integration is carried out using the triangulation laser scanner Konica Minolta Vivid 9i and the thermal camera Testo 882, but it can be exploited for generating 3D thermal models with a generic range sensor and a generic thermal camera. The two devices are kept independent during the acquisition phase, allowing the integration of 3D data and thermal data acquired at different times.

With regard to the extrinsic calibration, two methods are used, a more standard one relying on a manual selection of homologous points, and an "automatic" one, the latter based on finding the optimum of a particular function which evaluates the degree of filling of the reprojection of the 3D points inside the object silhouette in the thermogram. The former method is used to assess the effectiveness of the latter, which is proven to work well in the case study, but has room for improvements, especially in terms of robustness. Concerning the data fusion, we propose an easy to implement algorithm which is able to deal with complex object shapes, handle occlusions and cases of incomplete data from the range finder. Furthermore, the viewing angle is computed, and it is used to calculate a weight for each ray, in order to assign a proper temperature value in the zones in which, when integrating multiple thermograms, overlaps occur. It was shown how this can effectively reduce the error in the temperature due to the dependence of the emissivity on the viewing angle.

The integration methodology was first tested on a 3D-printed object and was then applied to a cultural heritage case study, and the results suggest that this approach can be effective and useful with an eye to integration and restoration. We are planning further tests to better investigate the effectiveness of the method.

Supplementary Materials: The following are available online at http://www.mdpi.com/2076-3417/10/3/828/s1.

Author Contributions: The authors have given an equal contribution to all the different parts of this article. All authors have read and agreed to the published version of the manuscript.

Funding: This research received no external funding.

Acknowledgments: The authors would like to thank Ing. Massimiliano Fantini (Romagna Tech S.C.P.A.) for his important contribution to 3D laser scanner survey about the statue of Agostino di Duccio.

Conflicts of Interest: The authors declare no conflict of interest.

References

1. Vollmer, M.; Möllmann, K.-P. *Infrared Thermal Imaging*; WILEY-VCH Verlag GmbH & Co. KGaA: Weinheim, Germany, 2010.
2. Soldofieri, F.; Dumoulin, J. Infrared thermography: From sensing principle to nondestructive testing considerations. In *Sensing the Past. Geotechnologies and the Environment*; Masini, N., Soldovieri, F., Eds.; Springer: Cham, Switzerland, 2017; Chapter 11; pp. 233–255.
3. Historic England 2018. *3D Laser Scanning for Heritage: Advice and Guidance on the Use of Laser Scanning in Archaeology and Architecture*; Historic England: Swindon, UK, 2018; Available online: https://historicengland. org.uk/images-books/publications/3d-laser-scanning-heritage/heag155-3d-laser-scanning/ (accessed on 23 January 2020).
4. Chernov, G.; Chernov, V.; Barboza Flores, M. 3D Dynamic Thermography System for Biomedical Applications. In *Application of Infrared to Biomedical Sciences*; Ng, E.Y., Etehadtavakol, M., Eds.; Springer: Singapore, 2017; pp. 517–545. [CrossRef]
5. Ordonez Muller, A.; Kroll, A. Generating High Fidelity 3-D Thermograms with a Handheld Real-Time Thermal Imaging System. *IEEE Sens. J.* **2017**, *17*, 774–783. [CrossRef]
6. Cardone, G.; Ianiro, A.; dello Ioio, G.; Passaro, A. Temperature Maps Measurements on 3D Surfaces with Infrared Thermography. *Exp. Fluids* **2012**, *52*, 375–385. [CrossRef]
7. Ju, X.; Nebel, J.-C.; Siebert, J.P. 3D Thermography Imaging Standardization Technique for Inflammation Diagnosis. In Proceedings of the Photonics Asia 2004, Beijing, China, 8–12 November 2004; pp. 266–273. [CrossRef]
8. Skala, K.; Lipić, T.; Sović, I.; Grubišić, I. Dynamic Thermal Models for Human Body Dissipation. *Period. Biol.* **2015**, *117*, 167–176.
9. Vidas, S.; Moghadam, P. HeatWave: A Handheld 3D Thermography System for Energy Auditing. *Energy Build.* **2013**, *66*, 445–460. [CrossRef]
10. Rangel, J.; Soldan, S. 3D Thermal Imaging: Fusion of Thermography and Depth Cameras. In Proceedings of the 2014 International Conference on Quantitative InfraRed Thermography, Bordeaux, France, 7–11 July 2014. [CrossRef]
11. Lussier, J.T.; Thrun, S. Automatic Calibration of RGBD and Thermal Cameras. In Proceedings of the 2014 IEEE/RSJ International Conference on Intelligent Robots and Systems, Chicago, IL, USA, 14–18 September 2014; pp. 451–458. [CrossRef]
12. Cao, Y.; Xu, B.; Ye, Z.; Yang, J.; Cao, Y.; Tisse, C.-L.; Li, X. Depth and Thermal Sensor Fusion to Enhance 3D Thermographic Reconstruction. *Opt. Express* **2018**, *26*, 8179. [CrossRef] [PubMed]
13. Yang, M.-D.; Su, T.-C.; Lin, H.-Y. Fusion of Infrared Thermal Image and Visible Image for 3D Thermal Model Reconstruction Using Smartphone Sensors. *Sensors* **2018**, *18*, 2003. [CrossRef] [PubMed]
14. Borrmann, D.; Nüchter, A.; Đakulović, M.; Maurović, I.; Petrović, I.; Osmanković, D.; Velagić, J. A Mobile Robot Based System for Fully Automated Thermal 3D Mapping. *Adv. Eng. Inf.* **2014**, *28*, 425–440. [CrossRef]
15. Seccia, L.; Iannucci, A.M.; Santopuoli, M.; Vernia, B. Early Pixels. *Image Process.* **1996**, *8*, 10–12.

16. Rizzi, A.; Voltolini, F.; Girardi, S.; Gonzo, L.; Remondino, F. Digital Preservation, Documentation and Analysis of Paintings, Monuments and Large Cultural Heritage with Infrared Technology, Digital Cameras and Range Sensors. In Proceedings of the XXI International CIPA Symposium, Athens, Greece, 1–6 October 2007.

17. Gianinetto, M.; Giussani, A.; Roncoroni, F.; Scaioni, M.; Di Milano, P.; Rilevamento, D.; Di Lecco, P.R. Integration of Multi-Source Close-Range Data. In Proceedings of the CIPA XX International Symposium, Torino, Italy, 26 September–1 October 2005.

18. Scaioni, M.; Rosina, E.; L'Erario, A.; Dìaz-Vilariño, L. Integration of Infrared Thermography and Photogrammetric Surveying of Built Landscape. *ISPRS Int. Arch. Photogramm. Remote. Sens. Spat. Inf. Sci.* **2017**, 153–160. [CrossRef]

19. Stanco, F.; Battiato, S.; Gallo, G. *Digital Imaging for Cultural Heritage Preservation*; CRC Press: Boca Raton, FL, USA, 2011.

20. Kawasue, K.; Win, K.D.; Yoshida, K.; Tokunaga, T. Black Cattle Body Shape and Temperature Measurement Using Thermography and KINECT Sensor. *Artif. Life Robot.* **2017**, *22*, 464–470. [CrossRef]

21. Robinson, D.W.; Simpson, R.; Parian, J.A.; Cozzani, A.; Casarosa, G.; Sablerolle, S.; Ertel, H. 3D Thermography for Improving Temperature Measurements in Thermal Vacuum Testing. *CEAS Space J.* **2017**, *9*, 333–350. [CrossRef]

22. Krefer, A.G.; Lie, M.M.I.; Borba, G.B.; Gamba, H.R.; Lavarda, M.D.; De Souza, M.A. A method for generating 3D thermal models with decoupled acquisition. *Comput. Methods Programs Biomed.* **2017**, *151*, 79–90. [CrossRef] [PubMed]

23. Jähne, B. *Practical Handbook on Image Processing for Scientific Applications*; CRC Press: Boca Raton, FL, USA, 1997.

24. Zhang, Z. A flexible new technique for camera calibration. *IEEE Trans. Pattern Anal. Mach. Intell.* **2000**, *22*, 1330–1334. [CrossRef]

25. Moghadam, P.; Bosse, M.; Zlot, R. Line-Based Extrinsic Calibration of Range and Image Sensors. In Proceedings of the 2013 IEEE International Conference on Robotics and Automation, Karlsruhe, Germany, 6–10 May 2013; pp. 3685–3691. [CrossRef]

26. Pandey, G.; Mcbride, J.R.; Savarese, S.; Eustice, R.M. Automatic Targetless Extrinsic Calibration of a 3D Lidar and Camera by Maximizing Mutual Information. In Proceedings of the Twenty-Sixth AAAI Conference on Artificial Intelligence, Toronto, Canada, 22–26 July 2012; pp. 2053–2059.

27. Scaramuzza, D.; Harati, A.; Siegwart, R. Extrinsic Self Calibration of a Camera and a 3D Laser Range Finder from Natural Scenes. In Proceedings of the 2007 IEEE/RSJ International Conference on Intelligent Robots and Systems, San Diego, CA, USA, 29 October–2 November 2007; pp. 4164–4169. [CrossRef]

28. Vidas, S.; Lakemond, R.; Denman, S.; Fookes, C.; Sridharan, S.; Wark, T. A Mask-Based Approach for the Geometric Calibration of Thermal-Infrared Cameras. *IEEE Trans. Instrum. Meas.* **2012**, *61*, 1625–1635. [CrossRef]

29. Gao, X.-S.; Hou, X.-R.; Tang, J.; Cheng, H.-F. Complete Solution Classification for the Perspective-Three-Point Problem. *IEEE Trans. Pattern Anal. Mach. Intell.* **2003**, *25*, 930–943. [CrossRef]

30. Reinbacher, C.; Ruther, M.; Bischof, H. Pose Estimation of Known Objects by Efficient Silhouette Matching. In Proceedings of the 2010 20th International Conference on Pattern Recognition, Istanbul, Turkey, 23–26 August 2010; pp. 1080–1083. [CrossRef]

31. Boyer, E. On Using Silhouettes for Camera Calibration. In *Computer Vision—ACCV 2006*; Narayanan, P.J., Nayar, S.K., Shum, H.-Y., Eds.; Springer: Berlin/Heidelberg, Germany, 2006; Volume 3851, pp. 1–10. [CrossRef]

32. Tateno, K.; Tombari, F.; Navab, N. When 2.5D Is Not Enough: Simultaneous Reconstruction, Segmentation and Recognition on Dense SLAM. In Proceedings of the 2016 IEEE International Conference on Robotics and Automation (ICRA), Stockholm, Sweden, 16–21 May 2016; pp. 2295–2302. [CrossRef]

33. Vidas, S.; Moghadam, P.; Sridharan, S. Real-Time Mobile 3D Temperature Mapping. *IEEE Sens. J.* **2015**, *15*, 1145–1152. [CrossRef]

Article

Effect of Hot Water Setting Temperature on Performance of Solar Absorption-Subcooled Compression Hybrid Cooling Systems

Jinfang Zhang [1,2,3], Zeyu Li [1,2,3,*], Hongkai Chen [1,2,3] and Yongrui Xu [1,2,3]

[1] School of Electric Power, South China University of Technology, Guangzhou 510640, China; 201720112923@mail.scut.edu.cn (J.Z.); gdchenhongkai@outlook.com (H.C.); xyrxuyongrui@163.com (Y.X.)

[2] Guangdong Province Key Laboratory of High Efficient and Clean Energy Utilization, South China University of Technology, Guangzhou 510640, China

[3] Guangdong Province Engineering Research Center of High Efficient and Low Pollution Energy Conversion, Guangzhou 510640, China

* Correspondence: epzeyuli@scut.edu.cn

Received: 6 October 2019; Accepted: 25 December 2019; Published: 29 December 2019

Abstract: The solar absorption-subcooled compression hybrid cooling system (SASCHCS) displays outstanding advantages in high-rise buildings. Since the performance coupling of collectors and absorption subsystems is stronger due to the absence of backup heat and the effect of generator setting temperature has not been realized adequately, it is highly important to study the relationship of SASCHCS operation and the set point temperature of hot water to prevent performance deterioration by inappropriate settings. Therefore, the paper mainly deals with the effect of collector and generator setting temperature. The investigation was based on the entire cooling period of a typical high-rise office building in subtropical Guangzhou. The off-design model of hybrid systems was built at first. Subsequently, the impact mechanism of setting temperature in two hot water cycles on facility operation was analyzed. It was found that the excessive rise of collector setting temperature deteriorated the energy saving, while the appropriate improvement of generator set point temperature was beneficial for the solar cooling. Besides, global optimization by the genetic algorithm displayed that 71.6 °C for the collector setting temperature with 64.5 °C for the generator was optimal for annual operation. The paper is helpful in enhancing the operation performance of SASCHCS.

Keywords: setting temperature of hot water; solar cooling; absorption chiller; subcooled compression; hybrid system

---

1. Introduction

The energy consumption of air conditioning accounts for 50% of building consumption and 15% of total electricity consumption, respectively [1,2]. Because solar energy is abundant and coincident with cooling loads, solar refrigeration systems display significant energy saving potential in air conditioning. Solar LiBr/H_2O single effect absorption chillers are widespread due to simplicity, high efficiency, as well as low cost among all solar thermal cooling [3]. However, it faces an economical obstacle in high-rise buildings owing to the excessive consumption of auxiliary thermal energy [4]. Considering that the number of high-rise buildings and the corresponding consumption grow notably because of increasing urban population and land price [5], more attention should be paid to the energy saving of such buildings. Li [6] proposed the solar absorption-subcooled compression hybrid cooling system (SASCHCS) as the economically feasible solution of high-rise buildings. It was found that the financial improvement mainly lies in the high performance of the absorption subsystem and the remarkable reduction of operational cost [7]. Besides, the prototype experiment showed that compressor works are saved by 22.2% on sunny days [8]. Therefore, the economic performance of SASCHCS rises dramatically, i.e., the payback period of hybrid systems is close to that of solar photovoltaic cooling (the most economical solution of solar refrigeration in recent years) in high-rise buildings [9]. It was shown that the area of compression subsystem evaporators and condensers as well as the size of absorption subsystems are critical for the exergoeconomic design of SASCHCS [10]. In particular, the optimal design of the above-mentioned parameters is dependent on the working condition distribution, i.e., absorption subsystems with moderate size are appropriate for the performance improvement if the hot water temperature mainly lies in the range 75–90 °C [11]. In addition to the system design, the performance also relies on the reasonable control of solar heat. The variable flow rate strategy in which flow rates of hot water are adjusted automatically according to the setting temperature is one of the widely-used control approaches in solar thermal systems. For the actual operation of SASCHCS, the performance suffers from notable decrease as a result of inappropriate setting temperature, i.e., the heat loss of collectors goes up and the useful heat drops dramatically by the increased temperature in spite of the rise in the coefficient of performance (COP) in absorption subsystems if the temperature of the generator hot water is set too high. Consequently, the relationship of SASCHCS operation and hot water set point temperature must be exact.

It was shown that the fixed outlet temperature is more appropriate than the constant inlet one for collector operation if flow rates of hot water can be controlled to make above-mentioned temperatures fixed during the entire period [12]. For solar domestic hot water facilities, it was found that the solar fraction associated with the constant temperature rise of collectors was only 2% more than that regarding fixed outlet temperatures [13]. Similar trends were also obtained by Rehman [14]. Araujo [15] compared the strategy with respect to constant and variable collector flow rates and found that the performance of solar domestic hot water systems with variable flow rate strategy was better than that with the constant one, especially for collectors with greater heat loss, i.e., the annual mean solar fraction with the variable flow rate strategy was 50% higher than that with the constant one for unglazed collectors. Furthermore, it was derived that the drop of set point temperature in collector outlet just enhanced 0.6% of heating power in solar district heating systems [16]. For solar power plants, it was concluded that the variable hot water flow rate strategy is more desirable than the constant one due to fewer parasitic loads [17]. In addition, it was shown that the setting temperature of the collector outlet relies on the tradeoff of power cycle efficiency and collector performance [18]. Camacho [19] found that the set point temperature of the collector outlet in winter should be 20 °C less than that in summer. Moreover, the operation of solar power plants becomes more flexible and performance rises remarkably when the collector setting temperature is allowed to change during the working period [20].

In addition to the solar heating and the power plant, the impact of collector setting temperature on performance of solar thermal cooling facilities has been researched as well. Qu [21] performed the comparison of solar LiBr/H_2O double effect absorption chillers with constant and variable hot water flow rates and concluded that systems with variable hot water flow rates extend the working period by 25%. Additionally, it was recommended that lowering the set point temperature of collectors as much as possible would be beneficial for the performance of solar absorption chillers [22]. Petela [23] found that the strategy in which the collector set point temperature was shifted from 160 °C to 140 °C in morning and afternoon nearly doubled the specific cooling power from 9:00–10:00 and from 15:00–16:00 compared with the strategy with a 160 °C setting temperature in collector outlet. It is noteworthy that the set point temperature of the collector outlet should not be less than 140 °C to avoid serious performance deterioration of solar NH_3/H_2O absorption chillers [24]. Similarly, it was shown that the strategy in which the collector setting temperature was adjusted in terms of the cooling load enhanced the solar fraction by 11% compared to the constant hot water flow rate strategy [25].

Although some valuable criteria regarding the variable hot water flow rate for solar thermal systems have been obtained, the strategy cannot be employed to guide the setting temperature of hot water for the SASCHCS adequately or precisely. Firstly, most of the studies with respect to the influence of collector setting temperature refer to facilities with backup heat. However, absorption subsystems of the SASCHCS are exclusively driven by solar energy, resulting in the stronger performance coupling of collectors and absorption subsystems. For instance, there are temperature rises of hot water in solar fields and generators through the heating of an auxiliary heater so that the performance of collectors/absorption chillers with generator/collector temperature is relatively weak. However, the hot water temperatures of collectors and generators are closely related to each other in the SASCHCS, i.e., the inlet temperature of generator hot water strongly relies on the collector outlet temperature, and the collector inlet temperature is mainly impacted by the stratification of storage tanks (the top layer temperature of storage tanks is equal to the inlet temperature of generator hot water) and the outlet temperature of generator hot water. Secondly, the effect of set point temperature in generator hot water has yet to be analyzed. As mentioned above, it is inferred that the setting temperature of generator hot water not only dramatically influences the COP of absorption chillers but is relevant to the collector efficiency. Motivated by the above-mentioned knowledge gaps, the relationship of set point temperature in hot water and SASCHCS performance was investigated thoroughly. The setting temperature of collectors and generators was taken into account. The research was based on the annual cooling period of typical high-rise office buildings in subtropical Guangzhou. The off-design model of SASCHCS was developed at first. Subsequently, the impact of setting temperature in two hot water cycles was precisely analyzed. Finally, the optimal set point temperature of hot water was derived through the genetic algorithm. The novelty of this paper is the illustration regarding the effect mechanism of hot water setting temperature as well as the presentation of corresponding working guidelines for the SASCHCS. The paper is helpful for the improvement of SASCHCS operation.

2. System Description: Operation Modes and Control Strategies

The schematic of SASCHCS is shown in Figure 1. The hybrid system mainly consists of thermal driving subsystems, absorption subsystems, and compression subsystems. The thermal driving subsystem is composed of the evacuated tube collector (ETC) and the stratified storage tank. Its working fluid is pressurized water. It is noted that the ETC is employed owing to the relatively high efficiency and the low cost. The absorption subsystem is a single effect $LiBr/H_2O$ absorption chiller, and the compression subsystem is a traditional vapor compression chiller with R410a. In the case of available solar energy, the cooling demand of buildings is met together by absorption and compression subsystems so that the compressor work is saved. When the solar heat is insufficient to feed absorption subsystems, the cooling load is fully covered by compression subsystems. Accordingly, the size of absorption subsystems is designed in terms of collector area, and that of compression subsystems is designed by the peak cooling demand of high-rise buildings. Moreover, there is no doubt that the reduction of compressor work relies on the amount of cooling output in absorption subsystems (subcooling power) and the corresponding conversion to the cooling power of compression subsystems. The expression of growth in the cooling capacity and the subcooling power was derived in our previous work [26]. It was displayed that the enhancement of cooling power to subcooling power is 0.9–1.2 when the superheating is fixed by the experiment [27].

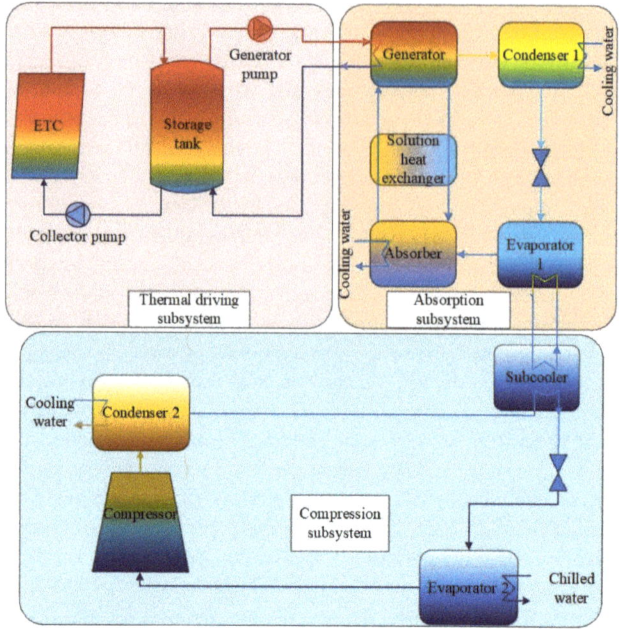

Figure 1. Schematic of solar absorption-subcooled compression hybrid cooling system (SASCHCS).

The control strategy of collector and generator pumps is demonstrated in Figure 2. The collector flow rate is controlled according to the setting temperature of the collector outlet, i.e., the flow rate of the collector hot water goes down automatically as the actual temperature of the collector outlet is lower than the set point one. The lowest collector flow rate is 10% of the nominal one to prevent thermal problems in the excessively low flow situation. Moreover, the collector pump is switched off when the temperature of the collector outlet is lower than the one of the storage tank bottom. It is restarted if the collector outlet temperature is 5 °C greater than the bottom temperature of the storage tank. In particular, the collector pump is turned off as the top layer temperature of storage tanks reaches 100 °C to avoid crystallization of absorption subsystems and is restarted as the temperature reaches less than 95 °C.

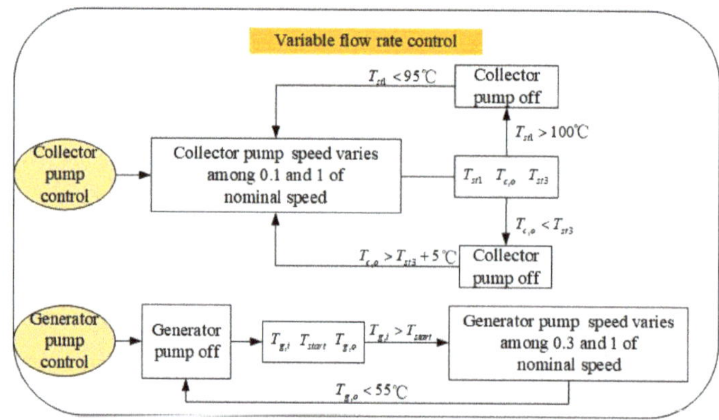

Figure 2. Control strategies for collector and generator pumps.

The generator pump is activated when the top layer temperature of storage tanks equals the activation temperature of absorption subsystems. The activation temperature is set to be 5 °C above the set point temperature of the generator hot water. Furthermore, the generator pump is turned off if the outlet temperature of the generator hot water falls to 55 °C because the temperature of hot water is too low to drive absorption subsystems from the experiment [8]. In another hand, the flow rate of the generator hot water is also controlled in terms of the setting temperature of the generator inlet. The minimal flow rate of the generator hot water is 30% of the rated one.

3. Model

The thermodynamic modeling of SASCHCS and model validation are described in this section.

3.1. Modeling of SASCHCS

The off-design model of SASCHCS was built based on the following assumption:

1. The system operates under quasi-steady state;
2. The cooling facility (absorption and compression subsystem) is adiabatic;
3. The evaporator and the condenser exit of compression subsystems are saturated;
4. The pressure loss of pipelines and heat exchangers is ignored;
5. The inlet hot water temperature of generator is equal to the top layer temperature of the storage tanks;
6. The inlet hot water of collectors is equal to the bottom layer temperature of storage tanks.

3.1.1. ETC and Storage Tank

The efficiency of ETC is calculated by the following expression [28]:

$$\eta_{etc,i} = Q_{etc,i}/AI_i = F_R[\tau\alpha - U_L(T_{etc,i} - T_{a,i})/I_i] \tag{1}$$

where F_R is the heat removal factor and equal to:

$$F_R = \frac{mc_p}{A_c U_L}\left[1 - \exp\left(-\frac{U_L F' A_c}{mc_p}\right)\right] \tag{2}$$

U_L is the overall heat loss coefficient and can be expressed by:

$$U_L = U_e + U_t \tag{3}$$

where U_e is header pipe edge loss coefficient and is expressed by:

$$U_e = \frac{2\pi\lambda_{ins}}{L_T \ln(D_{ext}/D_{int})} \tag{4}$$

U_t is the loss coefficient from the absorber tube to the ambient and can be defined as:

$$U_t = \frac{1}{\frac{1}{h_{g-a}} + \frac{1}{h_{p-g,r}+h_{p-g,c}}} \tag{5}$$

The collector efficiency factor, F', is derived by [29]:

$$F' = \frac{1/U_L}{L\left[\frac{1+U_L/C_b}{U_L(d+(L-d)F)} + \frac{1}{C_B} + \frac{1}{h_f\pi d}\right]} \tag{6}$$

Furthermore, the thermal efficiency of collectors based on the operational period is:

$$\eta_{etc} = \frac{\int_{\Delta t} Q_{etc,i} dt}{\int_{\Delta t} AI_i dt} \tag{7}$$

The storage tank has three layers, and the energy balance of each layer is calculated below:

$$\frac{1}{3}\rho V_{st} c_p \frac{dT_{st,i}}{dt} = m_c c_p (T_{st,i-1} - T_{st,i}) + m_g c_p (T_{st,i+1} - T_{st,i}) - h_{st} A_{st} (T_{st,i} - T_{a,i})/3 \tag{8}$$

The initial temperature of storage tank is equal to the surrounding temperature. It is stated that layer 1 donates the top layer of the storage tank and layer three is the bottom layer.

3.1.2. Absorption Subsystem and Compression Subsystem

The off-design model of cooling subsystems (the absorption subsystem and the compression one) was developed in our previous study [30]. The model of the absorption subsystem was based on the characteristic equation.

The relationship of the generator heat transfer and the generator hot water flow rate can be expressed as [31]:

$$UA = \left(\frac{m_g}{m_{g,rated}}\right)^{0.8} (UA)_{rated} \tag{9}$$

The cooling output of absorption subsystems is expressed as:

$$Q_{e,as} = s \cdot \Delta\Delta t - \alpha \cdot Q_{loss,as} = s \cdot \Delta\Delta t - s \cdot \Delta\Delta t_{min} \tag{10}$$

The parameter s is related to UA (multiplication of overall heat transfer coefficient and area in heat exchangers), and $\Delta\Delta t$ is relevant to mean temperature of hot water, cooling water, and chilled water. $\Delta\Delta t_{min}$ is calculated as follows [32]:

$$\Delta\Delta t_{min} = 1.9 + 0.1\Delta\Delta t \tag{11}$$

The COP of absorption subsystems is:

$$COP_{as} = \frac{\int Q_{e,as,i} dt}{\int Q_{g,i} dt} \tag{12}$$

The off-design model of compressions is based on the lumped parameter method. As a result, the thermodynamic characteristic of its component is controlled by mass and energy conservation:

$$\sum m_i = \sum m_o \tag{13}$$

$$Q = \sum m_i h_i - \sum m_o h_o = m c_p \Delta T = h \cdot A \cdot LMTD \tag{14}$$

The compressor work consumed by the SASCHCS is calculated by:

$$W = m_{cs}(h_{dis,s} - h_{suc})/\eta_s \tag{15}$$

Compared to the reference case (the cooling load is solely met by the compression subsystem), the energy saving of SASCHCS is:

$$E_{save} = \int_{\Delta t} (W_{ref,i} - W_i) dt \tag{16}$$

The COP of compression subsystems is:

$$COP_{cs} = \frac{\int Q_{e,cs,i}dt}{\int W_i dt} \tag{17}$$

3.2. Model Validation and Case Study

The ETC model was verified through the experimental data of Ghoneim [28]. A good agreement is displayed in Figure 3. The maximal deviation was 5.9%. In addition, the model validation of absorption subsystems and entire systems was implemented in our previous investigation by experiment [30]. As shown in Figure 4, the deviation regarding the model of absorption subsystems was within 10%. Additionally, it was demonstrated that the maximum relative error with respect to the model of entire systems was 3.58%, as shown in Table 1.

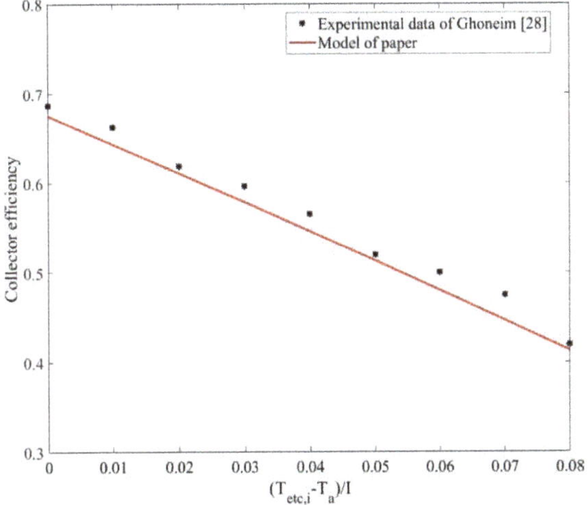

Figure 3. Validation of evacuated tube collector (ETC) model.

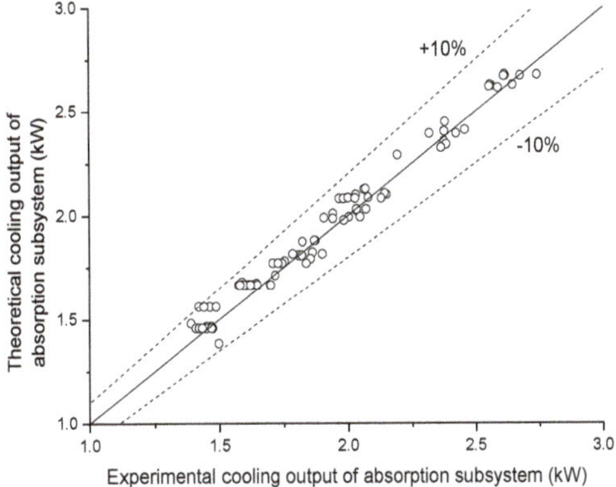

Figure 4. Model validation of absorption subsystem.

Table 1. Model validation of compression subsystem.

Parameter	Value		
	Model	**Experiment**	**Relative Error**
$T_{cw,cs,o}$ (°C)	31.53	32.7	3.58%
$T_{chw,cs,o}$ (°C)	10	9.9	1.01%
T_{dis} (°C)	59.68	59.4	0.47%
ΔT_{sub} (°C)	14.28	14.1	1.28%

A case study was performed to assess the relationship of hot water setting temperature and the performance of SASCHCS. A typical high-rise office building located in subtropical Guangzhou was considered. Such buildings consist of ten floors, and the total area of the chosen building is 3840 m² (area of each floor is 384 m²). As shown in Figure 5, the length and the width of every floor are 24 m and 16 m, respectively. There are ten rooms (seven offices, one meeting room, one rest room, as well as one machine room) on each story. It is noted that the machine room, the lift, and the staircase are excluded for cooling supply. Corresponding parameters of the above-mentioned building are listed in Table 2. The SASCHCS was employed to fulfill the cooling demand of office buildings. Design and operation parameters of SASCHCS are exhibited in Table 3. It is noted that the ETC was placed in the roof, and its installation area was designed to reasonably prevent the interference of each collector. Similar to the reference [33], parameters of absorption and compression subsystems were proportional to its nominal cooling capacity based on the size of our prototypes [8]. The typical monthly solar irradiance, the surrounding temperature, and the average data of annual measurement in our previous study [34] are exhibited in Figures 6 and 7, respectively. The typical monthly data are the mean of monthly measurement data, i.e., the solar irradiance of 8:00 in August is the average of 8:00 data from 1 to 31 August. Consequently, such typical meteorological data can reflect the solar irradiance and the ambient temperature of the entire month more reasonably. The above-mentioned data were recorded by a small wireless weather station model named DAVIS Vantage Pro 2. Furthermore, the data were verified by the meteorological information center and the maximal deviation was less than 10%. As displayed in Figure 8, the cooling load of buildings was calculated by the software DeST [35] (DeST is the building energy consumption analysis software developed by Tsinghua University). It is the free software package that simulates the building environment and HVAC (heating, ventilation and air conditioning) systems. DeST platform is based on more than 10 years of research data by the Institute of Environment and Equipment, Department of Building Science and Technology, Tsinghua University. The model of entire facilities is solved in the MATLAB environment [36] with 1 min time steps. The thermodynamic property of working fluid and refrigerant was obtained by Refprop 9 [37]. In particular, the cooling demand of high-rise buildings exclusively offered by compression subsystems served as the reference case in the calculation of energy saving regarding the SASCHCS.

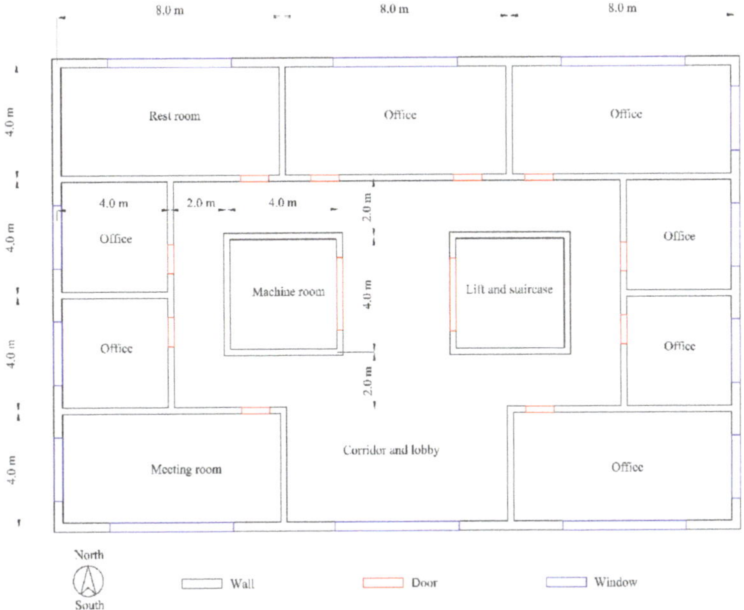

Figure 5. Layout and orientation of building.

Table 2. Load simulation assumptions and schedules for the case study.

Parameter	Value
Floor area	$384\ m^2$
Floor height	3.3 m
Floor number	10
Total area	$3840\ m^2$
Cooling season	From April to October
Air conditioning operation period	8:00–18:00
Window-to-wall ratio	0.5
Space temperature control	22–26 °C
Space humidity control	40–60%
Occupant density in office/meeting room/rest room/corridor and lobby	$0.1/0.3/0.3/0.2\ person/m^2$
Sensible heat load regarding people in office/meeting room/rest room/corridor and lobby	66/61/62/58 W/person
Humidity load regarding people in office/meeting room/rest room/corridor and lobby	0.102/0.109/0.068/0.184 kg/(Hr·person)
Lighting power density	$9\ W/m^2$
Electrical equipment power density in office/meeting room/rest room/corridor and lobby	$18/11/5/11\ W/m^2$
Ventilation rate	1 vol/h
Heat transfer coefficients of walls/windows/roof	$1.081/2.7/0.812\ W/(m^2K)$

Table 3. Operation parameters of SASCHCS.

Parameters	Unit	Value
ETC→		
Aperture area	m^2	270
Outer diameter of absorber tube	m	0.037
Thickness of absorber tube	m	0.0012
Outer diameter of glass tube	m	0.047
Thickness of glass tube	m	0.0012
Outer diameter of U-tube	m	0.008
Thickness of air layer	m	0.001
Thickness of copper fin	m	0.0006
Length of the header pipe per one U-tube	m	1.2
Nominal hot water flow of collector	kg/s	2.78
Tilted angle		20
Storage tank		
Aspect ratio		3.5
Heat loss coefficient	W/(m^2 K)	0.83
Nominal hot water flow of generator	kg/s	2.78
Volume of storage tank	m^3	2.7
Absorption subsystem		
Flow rate of cooling water in condenser 1/absorber/subcooler	kg/s	2.5/2.2/2.9
Inlet temperature of cooling water	°C	32
Nominal coefficient of performance (COP)		0.73
Compression subsystem		
Flow rate of cooling water in condenser 2	kg/s	27.3
Inlet temperature of cooling water	°C	32
Inlet/outlet temperature of chilled water	°C	12/7
Isentropic efficiency of compressor		0.7
Nominal COP		4.26

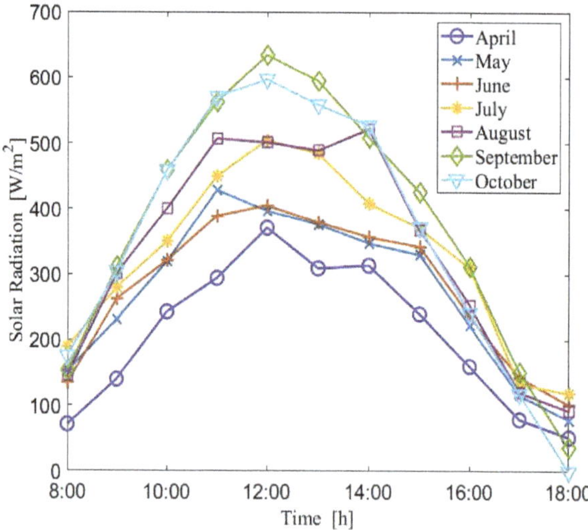

Figure 6. Solar radiation.

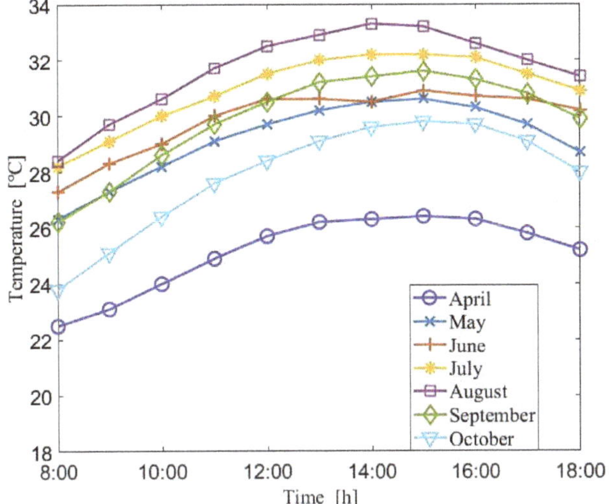

Figure 7. Ambient temperature.

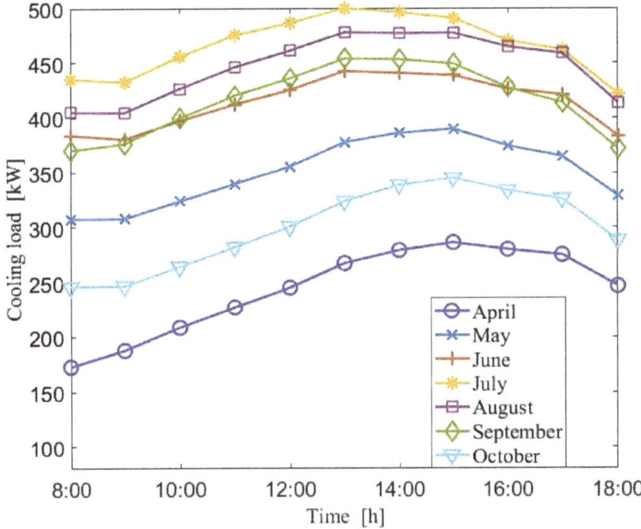

Figure 8. Cooling load.

4. Results and Discussion

This section includes two topics: (1) the impact of hot water setting temperature and (2) the optimization of set point temperature in two hot water cycles based on the annual period by the genetic algorithm. It is noteworthy that the analysis of hot water setting temperature was based on the August data. Furthermore, the influence of set point temperature in hot water cycles was analyzed step by step to show the exact relationship between hot water setting temperature and facility performance. Firstly, the quasi-steady variation of hot water temperature and flow rate for two set point temperatures of hot water was analyzed. Secondly, the useful heat of collectors, COP, and cooling output of absorption subsystems for different hot water setting temperatures was illustrated. Thirdly, the monthly energy savings of SASCHCS for different set point temperatures of hot water cycles was elucidated.

4.1. Effect of Hot Water Setting Temperature

The variation of hot water temperature and flow rate with two set point temperatures of collector outlet is demonstrated in Figure 9. It is noteworthy that the setting inlet temperature of generator hot water was 70 °C. For the case in which the set point temperature of the collector outlet was 75 °C, it was observed that the collector flow rate went up gradually and maintained the nominal one from 11:05 to 14:45. Subsequently, the collector flow rate came down gradually and maintained the minimal one owing to the drop of solar irradiance. It was seen that the collector pump stopped at 16:57 because the collector outlet temperature was less than the bottom temperature of the storage tanks. Simultaneously, the collector outlet temperature kept the setting one with a 2 °C increase from 9:33 to 16:02. It dropped to 58 °C quickly in the end of the operation. Additionally, the generator pump was activated at 11:07, and the flow rate of generator hot water kept the rated one until 15:27. Subsequently, the flow rate of generator hot water went down quickly to the lowest one, and the generator pump was switched off at 16:37 since the hot water temperature of the generator outlet was less than 55 °C. Moreover, the hot water temperature of the generator inlet remained at 75 °C until 14:35 and reduced to 64 °C gradually in the end of the operation. For the case in which the set point temperature of the collector outlet was 105 °C, the collector flow rate nearly held the minimal one during the entire period. Besides, the collector outlet temperature attained the set point one in midday. It was found that the duration that the collector outlet temperature maintained the setting one reduced by 57% compared to the case in which the set point temperature of the collector outlet was 75 °C. The trend of collector outlet temperature for two setting temperatures overlapped after 16:00. Furthermore, the generator pump was activated 39 min earlier due to the faster improvement of top layer temperature in the storage tank. The sudden fall of generator hot water flow rate at the start of the generator pump was mainly attributed to the insufficient heat of the storage tank. Subsequently, the flow rate of generator hot water gradually approached the rated one at 14:00 and then decreased quickly to the lowest one. Furthermore, the generator inlet temperature almost held the set point one until 15:10. The generator pump was turned off at 16:19, which was 18 min earlier than the case in which the set point temperature of the collector outlet was 75 °C.

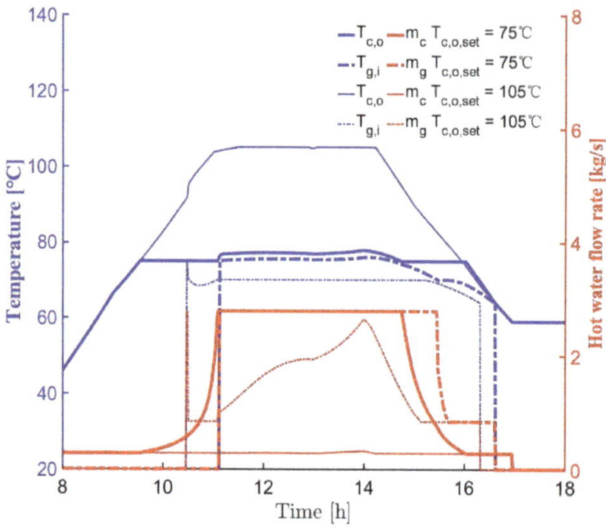

Figure 9. Hot water temperature and flow rate for different $T_{c,o,set}$.

The variation of hot water temperature and flow rate with two set point temperatures of generator inlet is displayed in Figure 10. It is noteworthy that the setting temperature of the collector outlet was 95 °C. It was seen that collector flow rates for two setting temperatures of the generator inlet nearly overlapped except for the midday. The collector flow rate corresponding to 90 °C of Tg, i, set became quadratic in this period owing to the strong solar irradiance and generator consumption. The collector outlet temperatures for two set point temperatures of the generator inlet were extremely similar except the duration when the collector outlet temperature kept the setting one for 90 °C of Tg, i, set and extended 15 min. The flow rate of generator hot water for 90 °C of Tg, i, set came down quickly after the activation of the generator pump. Similarly, its generator inlet temperature just maintained the set point one for 51 min, and the corresponding period was 20% of the one for 70 °C of Tg, i, set. In addition, it was shown that the activation and the stop of the generator pump were delayed by 127 min and 46 min, respectively, for the case in which the setting temperature of the generator inlet was 90 °C.

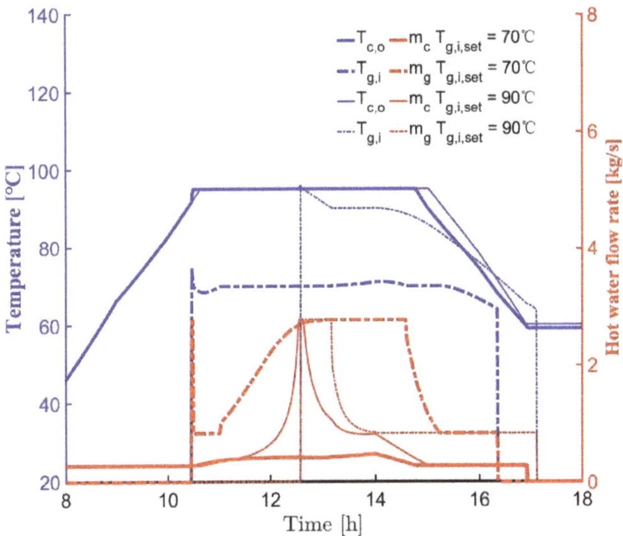

Figure 10. Hot water temperature and flow rate for different Tg, i, set.

The impact of collector setting temperature with 70 °C set temperature of generator hot water is demonstrated in Figure 11. It was shown that trends of monthly useful heat in collectors and cooling capacity of absorption subsystems were similar. Both grew slightly as the set point temperature of the collector rose to 80 °C at first. It is known that the improvement of collector setting temperature decreases the collector flow rate so that the consequent drop in collector inlet temperature is favorable compared to the lower heat loss of collectors. Nevertheless, the excessive increase of collector setting temperature seriously deteriorated the performance, i.e., the monthly useful heat of collectors and the cooling capacity of absorption subsystems came down by 9.3% and 11.6%, respectively, if the collector set point temperature went up from 80 °C to 105 °C. The significant decrease of heat transfer coefficient caused by the excessive drop of collector flow rate led to the above-mentioned phenomenon. In addition, the COP of absorption subsystems with the set point temperature of collectors was quadratic as well. It enhanced by 1.8% when the collector setting temperature grew from 80 °C to 105 °C, which was led by the increased operation period of absorption subsystems. In general, the enhancement of the absorption subsystem COP was offset by the reduction of collector useful heat so that the excessive improvement of collector set point temperature was adverse for the solar cooling.

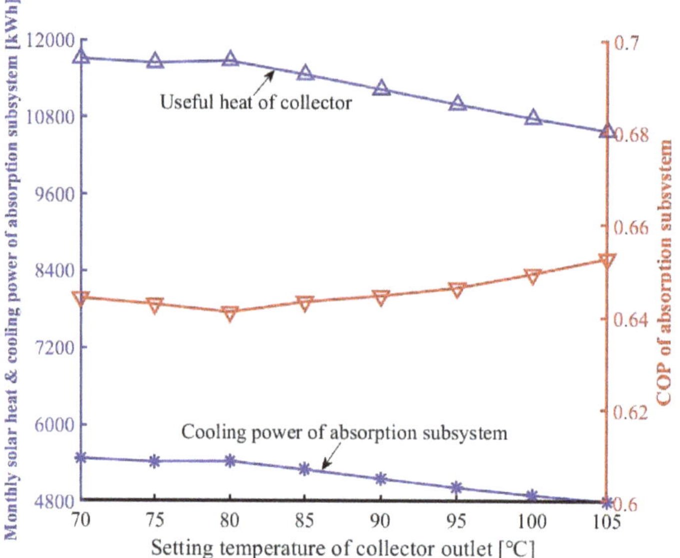

Figure 11. Impact of setting temperature of collector outlet.

The effect of generator setting temperature with 95 °C collector set point temperature is displayed in Figure 12. It was seen that the rise of generator set point temperature decreased the solar heat, i.e., useful heat of the collector went down by 11.8% as the setting temperature of the generators went up by 30 °C. This was attributed to the enhanced generator set point temperature increasing the bottom layer temperature of the storage tanks. Thereby, the rise of collector inlet temperature lowered the amount of solar heat. However, the COP of absorption subsystem grew with the enhancement of generator setting temperature except when the set point temperature of the generator exceeded 85 °C. It was shown that the COP of the absorption subsystem rose by 10.1% as the generator setting temperature went up from 60 °C to 85 °C. The above-mentioned phenomenon was attributed to the influence of increased generator hot water temperature surpassing the one of decreased flow rate. Accordingly, there was an optimal setting temperature of the generator that maximized the cooling power of absorption subsystems. It was derived that the optimal generator set point temperature was 75 °C when the setting temperature of the collector outlet was 95 °C. Additionally, the cooling output of absorption subsystems with 75 °C generator setting temperature was 13.6% more than that in the 60 °C one. In general, appropriate improvement of generator set point temperature was beneficial for the solar cooling though the amount of solar heat went down slightly.

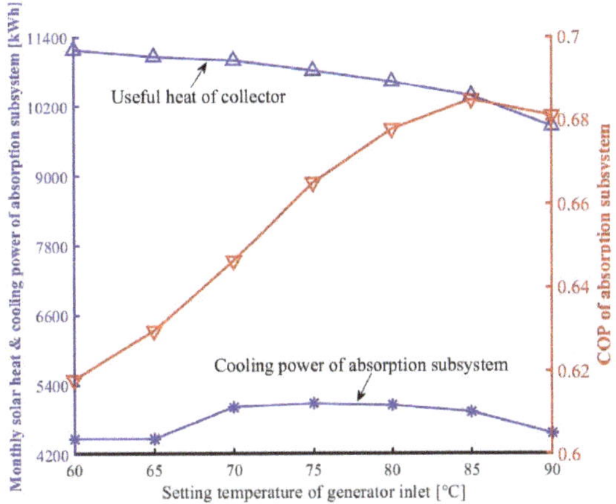

Figure 12. Impact of set point temperature of generator inlet.

The monthly energy savings of SASCHCS for different setting temperatures of the collector outlet are shown in Figure 13. It was noted that the set point temperature of the generator was 70 °C. As expected, it was observed that the higher the cooling capacity of absorption subsystems was, the higher the energy saving of hybrid systems became from the August data. It was demonstrated that qualitative trends of energy saving with collector set point temperature based on different monthly data were similar. Furthermore, the energy saving trends were independent from the set point temperature of the collector as the collector setting temperature was higher than 85 °C for April, May, and June (months with low and moderate solar irradiance). This was illustrated by the fact that the hot water could not be heated to such excessively high temperatures by the weak solar irradiance, and even its flow rate was reduced to the minimal one. This also implied that the set point temperature of the collector outlet below 85 °C was effective for the hot water control from April to June. Moreover, the excessive enhancement of the collector setting temperature lowered the performance of SASCHCS, i.e., the monthly energy savings of May and June only fell by 7.6% and 7.9%, respectively. The above-mentioned effect became notable for the months with strong solar irradiance. For example, the monthly energy savings of July, August, September, and October came down by 15%, 10.7%, 10.1%, and 11%, respectively, if the collector setting temperature grew from 80 °C to 105 °C.

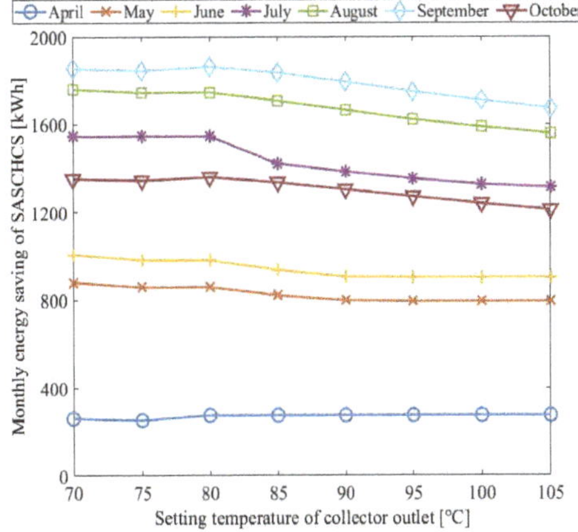

Figure 13. Energy saving for different collector setting temperatures.

The monthly energy savings of SASCHCS for different setting temperatures of the generator inlet are exhibited in Figure 14. It is noteworthy that the collector set point temperature was 95 °C. The energy savings with set point temperature of the generator were quadratic in the entire period. In addition, it was found that the optimal setting temperature of generators was around 70–75 °C. The effect of the generator set point temperature on the performance was stronger for the months with weak and moderate solar irradiance, i.e., the monthly energy savings of April, May, June, July, August, September, and October rose by 123.7%, 38.7%, 38.2%, 25%, 12.3%, 8.8%, and 10.6%, respectively, when the setting temperature of generators went up from 60 °C to the optimal one. This was attributed to the rise of absorption subsystem COP by the increased generator hot water temperature dominating the performance of SASCHCS in April to June. In particular, the excessively high setting temperature of the generator deteriorated the solar cooling dramatically in April to June. The reason was that such a relatively high set point temperature of the generator was extremely difficult to reach by the weak solar irradiance, thus the duration of absorption subsystems went down dramatically.

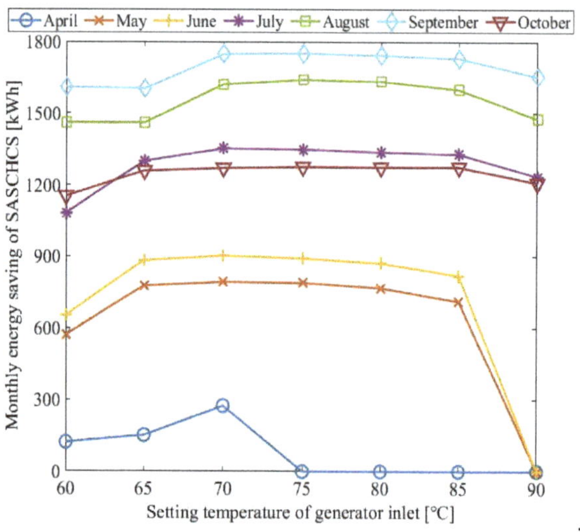

Figure 14. Energy saving for different generator setting temperatures.

The optimal collector setting temperature for different set point temperatures of the generator is listed at Table 4. It was shown that the optimal collector setting temperature associated with the certain set point temperature of generators was same regardless of month. Therefore, it can be said that the optimal setting temperature of the collector outlet was independent from the meteorological data. Furthermore, the optimal collector set point temperature strongly relied on the setting temperature of the generator, i.e., it was around 8–10 °C above the generator set point temperature. This was explained by the fact that the stratification of storage tanks, the critical factor of solar cooling, was kept by the synchronized growth of collector setting temperature. It should be noted that the amount of useful heat in collectors was subjected to remarkable reduction through the excessive rise of collector set point temperature.

Table 4. Optimal $T_{c,o,set}$ for different $T_{g,i,set}$.

$T_{g,i,set}$ (°C)	Optimal Collector Outlet Set Temperature (°C)						
	Apr	May	Jun	Jul	Aug	Sept	Oct
65	75	75	75	75	75	75	75
75	-	83	83	83	83	84	83
85	-	94	94	94	94	95	94

4.2. Global Optimization of Setting Temperature in Two Hot Water Cycles

According to the analysis of Section 4.1, it is known that the appropriate increase of generator setting temperature is favorable for the COP of absorption subsystems. However, the improvement of generator set point temperature depends on the rise of collector setting temperature leading to the decrease of solar heat. Consequently, the global optimization of hot water set point temperature is extremely essential. Moreover, because the optimal setting temperature of hot water was shown to be independent from the meteorological data, the result of global optimization is convenient for the operation. Such optimization was done by the genetic algorithm function in the MATLAB environment. The maximum generation number and the population number were set as 50 and 20, respectively. The ranges of the collector and the generator setting temperature were 70–105 °C and 60–90 °C, respectively. Additionally, the maximal annual energy savings was employed as the objective function.

The annual performance of SASCHCS in the optimal case is shown in Table 5. It was shown that 71.6 °C of the collector setting temperature with 64.5 °C of the generator one was optimal for the annual operation of SASCHCS. The corresponding peak energy savings of SASCHCS were 8841.3 kWh/year, which was equivalent to 32.75 kWh/m^2 of a specific one. Additionally, the annual collector efficiencies, the COP of absorption and compression subsystems, were 0.39, 0.63, and 4.86, respectively, in the optimal case. It was derived that the COP of compression subsystems enhanced by 14.1% due to the solar cooling compared to its nominal COP.

Table 5. Annual performance of SASCHCS in the optimal case.

Q_{etc} (kWh)	$Q_{e,as}$ (kWh)	E_{save} (kWh)	η_{etc}	COP_{as}	COP_{cs}
73310.7	30854.7	8841.3	0.39	0.63	4.86

5. Conclusions

The influence of hot water setting temperature on the operation and the performance of SASCHCS was analyzed in detail based on the data of the entire cooling period in subtropical Guangzhou. The corresponding conclusions are summarized as follows:

1. Despite the fact that the COP of the absorption subsystem went up slightly with the increased hot water temperature, the excessive improvement of the collector setting temperature was harmful for the energy saving, since the serious reduction of heat transfer caused by the low flow rate notably decreased the amount of collector useful heat. It was derived that the daily cooling output of the absorption subsystem based on August data dropped by 11.6% as the collector set point temperature went up from 80 °C to 105 °C.

2. Although the enhanced temperature of generator hot water lowered the amount of solar heat, the appropriate rise of generator set point temperature was favorable for the solar cooling owing to the remarkable growth of COP in absorption subsystems, i.e., the daily cooling capacity of the absorption subsystem based on August data enhanced by 13.6% if the setting temperature of the generator grew from 60 °C to 75 °C.

3. The collector set point temperature should be 8–10 °C above the generator one in terms of the tradeoff of collector useful heat and absorption subsystem COP. In particular, the above-mentioned relationship is independent from the meteorological data.

4. It was demonstrated that a 71.6 °C collector setting temperature with a 64.5 °C generator one was optimal for the annual operation by the global optimization. The corresponding peak annual specific energy saving of SASCHCS was 32.75 kWh/m^2. In addition, the annual collector efficiencies, the COP of absorption and compression subsystems, were 0.39, 0.63, and 4.86, respectively, in the optimal case.

Author Contributions: J.Z. contributes to data analysis and writing. Z.L. contributes to conceptualization and methodology. H.C. contributes to formal analysis. Y.X. contributes to the validation. All authors have read and agreed to the published version of the manuscript.

Funding: This work is supported by: (1) Natural Science Foundation of Guangdong Province under the contract No. 2018A030313310, (2) Guangzhou Municipal Science and Technology Project under the contract No. 201904010218, (3) State Key Laboratory of Compressor Technology under the contract No. SKL-YSJ201806, (4) Key Laboratory of Efficient and Clean Energy Utilization of Guangdong Higher Education Institutes under the contract No. KLB10004.

Conflicts of Interest: The authors declare no conflicts of interest.

Nomenclature

Symbol	Description	Symbol	Description
A	area (m^2)	V	volume (m^3)
A_c	the outer surface area of absorber tube (m^2)	W	compressor work (kW)
COP	coefficient of performance	ΔT	temperature difference (°C)
C_b	synthetic conductance W/(mK)	$\Delta\Delta t$	double difference of temperatures (°C)
C_B	bond conductance W/(mK)		*Greek symbols*
c_p	specific heat (kJ/kgK)	α	absorptivity, distribution UA parameter
d	diameter of the U-tube (m)	η	efficiency
D_{ext}	external diameter of insulation (m)	λ_{ins}	header pipe insulation thermal conductivity (W/mK)
D_{int}	internal diameter of insulation (m)	π	pi
E_{save}	energy saving (kWh)	ρ	density (kg/m^3)
F	fin efficiency of straight fin collector efficient factor	τ	transmissivity
F'	heat removal factor		*Subscripts*
F_R	heat transfer coefficient (W/m^{2}K)	a	ambient
h	specific enthalpy (kJ/kg)	as	absorption subsystem
h_f	convection heat transfer coefficient between the fluid and the U-tube W/(m^{2}K)	avg	average
h_{g-a}	convection heat transfer coefficient W/(m^{2}K)	cs	compression subsystem
$h_{p-g,r}$	radiation heat transfer coefficient, W/(m^{2}K)	dis	discharge
$h_{p-g,c}$	heat transfer coefficient due to the conduction W/(m^{2}K)	e	evaporator
I	solar irradiance (W/m^2)	etc	evacuated tube collector
L	the circumferential distance between the U-tubes (m)	g	generator
$LMTD$	logarithmic mean temperature difference (°C)	i	inlet, current time
L_T	length of the header pipe per one U-tube (m)	$loss$	loss
m	mass flow rate (kg/s)	min	minimum
Q	heat load (kW)	o	outlet
t	time	$rated$	rated
T	temperature (°C)	ref	reference
UA	multiplication of heat transfer coefficient and area (W/K)	s	isentropic
U_e	the heat loss coefficient from absorber tube to ambient W/(m^{2}K)	suc	suction
U_L	overall loss coefficient W/(m^{2}K)	st	storage tank
U_t	the edge loss coefficient of the header tube W/(m^{2}K)		

References

1. Mohammed, O.W.; Yanling, G. Comprehensive Parametric Study of a Solar Absorption Refrigeration System to Lower Its Cut In/Off Temperature. *Energies* **2017**, *10*, 1746. [CrossRef]
2. Perez-Lombard, L.; Ortiz, J.; Pout, C. A review on buildings energy consumption information. *Energy Build.* **2008**, *40*, 394–398. [CrossRef]

3. Xu, Z.; Wang, R. Comparison of CPC driven solar absorption cooling systems with single, double and variable effect absorption chillers. *Sol. Energy* **2017**, *158*, 511–519. [CrossRef]

4. Noro, M.; Lazzarin, R. Solar cooling between thermal and photovoltaic: An energy and economic comparative study in the Mediterranean conditions. *Energy* **2014**, *73*, 453–464. [CrossRef]

5. Colmenar-Santos, A.; Vale-Vale, J.; Borge-Diez, D.; Requena-Pérez, R. Solar thermal systems for high rise buildings with high consumption demand: Case study for a 5 star hotel in Sao Paulo, Brazil. *Energy Build.* **2014**, *69*, 481–489. [CrossRef]

6. Li, Z.; Jing, Y.; Liu, J. Thermodynamic study of a novel solar LiBr/H2O absorption chiller. *Energy Build.* **2016**, *133*, 565–576. [CrossRef]

7. Xu, Y.; Jiang, N.; Pan, F.; Wang, Q.; Gao, Z.; Chen, G. Comparative study on two low-grade heat driven absorption-compression refrigeration cycles based on energy, exergy, economic and environmental (4E) analyses. *Energy Convers. Manag.* **2017**, *133*, 535–547. [CrossRef]

8. Yu, J.; Li, Z.; Chen, E.; Xu, Y.; Chen, H.; Wang, L. Experimental assessment of solar absorption-subcooled compression hybrid cooling system. *Sol. Energy* **2019**, *185*, 245–254. [CrossRef]

9. Li, Z.; Liu, L. Economic and environmental study of solar absorption-subcooled compression hybrid cooling system. *Int. J. Sustain. Energy* **2019**, *38*, 123–140. [CrossRef]

10. Li, Z.; Liu, L.; Jing, Y. Exergoeconomic analysis of solar absorption-subcooled compression hybrid cooling system. *Energy Convers. Manag.* **2017**, *144*, 205–216. [CrossRef]

11. Jing, Y.; Li, Z.; Chen, H.; Lu, S.; Lv, S. Exergoeconomic design criterion of solar absorption-subcooled compression hybrid cooling system based on the variable working conditions. *Energy Convers. Manag.* **2019**, *180*, 889–903. [CrossRef]

12. Zamfir, E.; Bădescu, V.; Badescu, V. Different strategies for operation of flat-plate solar collectors. *Energy* **1994**, *19*, 1245–1254. [CrossRef]

13. Csordas, G.; Brunger, A.; Hollands, K.; Lightstone, M. Plume entrainment effects in solar domestic hot water systems employing variable-flow-rate control strategies. *Sol. Energy* **1992**, *49*, 497–505. [CrossRef]

14. Rehman, H.U.; Hirvonen, J.; Sirén, K. Design of a Simple Control Strategy for a Community-size Solar Heating System with a Seasonal Storage. *Energy Procedia* **2016**, *91*, 486–495. [CrossRef]

15. Araújo, A.; Silva, R.; Pereira, V. Solar thermal modeling for rapid estimation of auxiliary energy requirements in domestic hot water production: On-off versus proportional flow rate control. *Sol. Energy* **2019**, *177*, 68–79. [CrossRef]

16. Bava, F.; Furbo, S. Impact of different improvement measures on the thermal performance of a solar collector field for district heating. *Energy* **2018**, *144*, 816–825. [CrossRef]

17. Tehrani, S.S.M.; Taylor, R.A. Off-design simulation and performance of molten salt cavity receivers in solar tower plants under realistic operational modes and control strategies. *Appl. Energy* **2016**, *179*, 698–715. [CrossRef]

18. Yilmazoglu, M.Z. Effects of the selection of heat transfer fluid and condenser type on the performance of a solar thermal power plant with technoeconomic approach. *Energy Convers. Manag.* **2016**, *111*, 271–278. [CrossRef]

19. Camacho, E.; Gallego, A. Optimal operation in solar trough plants: A case study. *Sol. Energy* **2013**, *95*, 106–117. [CrossRef]

20. Brodrick, P.G.; Brandt, A.R.; Durlofsky, L.J. Operational optimization of an integrated solar combined cycle under practical time-dependent constraints. *Energy* **2017**, *141*, 1569–1584. [CrossRef]

21. Qu, M.; Yin, H.; Archer, D.H. A solar thermal cooling and heating system for a building: Experimental and model based performance analysis and design. *Sol. Energy* **2010**, *84*, 166–182. [CrossRef]

22. Calise, F.; D'Accadia, M.D.; Vanoli, R. Dynamic simulation and parametric optimisation of a solar-assisted heating and cooling system. *Int. J. Ambient. Energy* **2010**, *31*, 173–194. [CrossRef]

23. Petela, K.; Manfrida, G.; Szlek, A. Advantages of variable driving temperature in solar absorption chiller. *Renew. Energy* **2017**, *114*, 716–724. [CrossRef]

24. Weber, C.; Berger, M.; Mehling, F.; Heinrich, A.; Núñez, T. Solar cooling with water–ammonia absorption chillers and concentrating solar collector—Operational experience. *Int. J. Refrig.* **2014**, *39*, 57–76. [CrossRef]

25. Shirazi, A.; Pintaldi, S.; White, S.D.; Morrison, G.L.; Rosengarten, G.; Taylor, R.A. Solar-assisted absorption air-conditioning systems in buildings: Control strategies and operational modes. *Appl. Therm. Eng.* **2016**, *92*, 246–260. [CrossRef]

26. Li, Z.; Chen, E.; Jing, Y.; Lv, S. Thermodynamic relationship of subcooling power and increase of cooling output in vapour compression chiller. *Energy Convers. Manag.* **2017**, *149*, 254–262. [CrossRef]

27. Chen, E.; Li, Z.; Yu, J.; Xu, Y.; Yu, Y. Experimental research of increased cooling output by dedicated subcooling. *Appl. Therm. Eng.* **2019**, *154*, 9–17. [CrossRef]

28. Ghoneim, A.A. Performance optimization of9 evacuated tube collector for solar cooling of a house in hot climate. *Int. J. Sustain. Energy* **2018**, *37*, 193–208. [CrossRef]

29. Ma, L.; Lu, Z.; Zhang, J.; Liang, R. Thermal performance analysis of the glass evacuated tube solar collector with U-tube. *Build. Environ.* **2010**, *45*, 1959–1967. [CrossRef]

30. Li, Z.; Yu, J.; Chen, E.; Jing, Y. Off-Design Modeling and Simulation of Solar Absorption-Subcooled Compression Hybrid Cooling System. *Appl. Sci.* **2018**, *8*, 2612. [CrossRef]

31. Miao, D. *Simulation Model of a Single-Stage Lithium Bromide-Water Absorption Unit*; NASA Technical Paper 1296; NASA: Washington, DC, USA, 1978.

32. Arnavat, M.P.; López-Villada, J.; Bruno, J.C.; Coronas, A. Analysis and parameter identification for characteristic equations of single- and double-effect absorption chillers by means of multivariable regression. *Int. J. Refrig.* **2010**, *33*, 70–78. [CrossRef]

33. Chen, H.; Li, Z.; Xu, Y. Evaluation and comparison of solar trigeneration systems based on photovoltaic thermal collectors for subtropical climates. *Energy Convers. Manag.* **2019**, *199*, 111959. [CrossRef]

34. Li, Z.; Ye, X.; Liu, J. Performance analysis of solar air cooled double effect LiBr/H2O absorption cooling system in subtropical city. *Energy Convers. Manag.* **2014**, *85*, 302–312. [CrossRef]

35. De ST. Available online: http://update.dest.com.cn/ (accessed on 12 October 2019).

36. MATLAB. Available online: https://ww2.mathworks.cn/ (accessed on 12 October 2019).

37. Refprop9. Available online: http://www.nist.gov/srd/nist23.cfm (accessed on 12 October 2019).

Article

Investigation on Roof Segmentation for 3D Building Reconstruction from Aerial LIDAR Point Clouds

Raffaele Albano

School of Engineering, University of Basilicata, 85100 Potenza, Italy; raffaele.albano@unibas.it

Received: 18 September 2019; Accepted: 31 October 2019; Published: 2 November 2019

Abstract: Three-dimensional (3D) reconstruction techniques are increasingly used to obtain 3D representations of buildings due to the broad range of applications for 3D city models related to sustainability, efficiency and resilience (i.e., energy demand estimation, estimation of the propagation of noise in an urban environment, routing and accessibility, flood or seismic damage assessment). With advancements in airborne laser scanning (ALS), 3D modeling of urban topography has increased its potential to automatize extraction of the characteristics of individual buildings. In 3D building modeling from light detection and ranging (LIDAR) point clouds, one major challenging issue is how to efficiently and accurately segment building regions and extract rooftop features. This study aims to present an investigation and critical comparison of two different fully automatic roof segmentation approaches for 3D building reconstruction. In particular, the paper presents and compares a cluster-based roof segmentation approach that uses (a) a fuzzy c-means clustering method refined through a density clustering and connectivity analysis, and (b) a region growing segmentation approach combined with random sample consensus (RANSAC) method. In addition, a robust 2.5D dual contouring method is utilized to deliver watertight 3D building modeling from the results of each proposed segmentation approach. The benchmark LIDAR point clouds and related reference data (generated by stereo plotting) of 58 buildings over downtown Toronto (Canada), made available to the scientific community by the International Society for Photogrammetry and Remote Sensing (ISPRS), have been used to evaluate the quality of the two proposed segmentation approaches by analysing the geometrical accuracy of the roof polygons. Moreover, the results of both approaches have been evaluated under different operating conditions against the real measurements (based on archive documentation and celerimetric surveys realized by a total station system) of a complex building located in the historical center of Matera (UNESCO world heritage site in southern Italy) that has been manually reconstructed in 3D via traditional Building Information Modeling (BIM) technique. The results demonstrate that both methods reach good performance metrics in terms of geometry accuracy. However, approach (b), based on region growing segmentation, exhibited slightly better performance but required greater computational time than the clustering-based approach.

Keywords: LIDAR point clouds; 3D urban model; rooftop modeling; segmentation; reconstruction; 3D building

1. Introduction

Significant acceleration of the urbanization rate is contributing to a persistent vertical expansion in many cities in order to meet the increasing demand for living and working space [1]. In this context, the three-dimensional form of a city has important implications for a city's sustainability, efficiency, and resilience [2]. Considering the expected urban expansion in the near future across the globe [3], understanding the relation between the three-dimensional urban form, critical infrastructure operations, and quality-of-life measures is one of the most important research and technical challenges. The growing interest in generating 3D city models is motivated by a broad range of applications, such as estimation of solar irradiation, energy demand estimation, classification of building types, visibility analysis, 3D cadastre, visualization for navigation, urban planning, emergency response, computation fluid dynamics, change detection, flooding, archaeology, forest management, and virtual tours [4].

Moreover, a decrease in the cost of remote sensing technology and data storage in recent years has contributed to the expansion of urban morphology studies [4,5]. In particular, the emergence of airborne laser, a leading technology for the extraction of information about physical surfaces, enables substantial advances of in-depth availability of data on buildings and infrastructure, as well as over large scales. By directly providing measurements of surface heights with high point density and high level of accuracy, light detection and ranging (LIDAR) technology can improve the automation level in accurate and efficient 3D reconstruction of urban models [6–9]. The 3D information about buildings and related structures that can be retrieved from the data acquired by airborne LIDAR is usually characterized by directness and simplicity. However, practice shows that the massive number of points requires introducing some level of organization into the data before extraction of information can become effective (e.g., aggregating points with similar features into segments in the form of surfaces) [10]. Therefore, producing 3D building reconstruction by manual or semi-automatic methods could be a very time consuming and challenging task. Hence, the generation of 3D building models in a simple and quick way is becoming attractive [11]. Indeed, in the last decade, the automatic 3D reconstruction of buildings from airborne data is an active area of research among photogrammetry, computer graphics, and remote sensing communities [3,5,12–14].

Among the great variety of reconstruction methods from airborne laser scanning (ALS) proposed in literature (see [3] for a complete review), the data-driven polyhedral method is one of the most commonly used rooftop modeling techniques and can be adapted for generating building models with both simple and complex roof topology [5]. Moreover, polyhedron buildings are quite common in urban areas [7]. These methods use a bottom-up approach that begins with the extraction of primitives (e.g., planes, cylinders, cones and spheres), followed by analysing primitive topology in 3D space, extracting and grouping them to form building models. In summary, the problem of building rooftop reconstruction based on a data-driven framework is transformed into a problem of consistency maintenance of topological relationships among rooftop primitives, primitive boundaries or their combinations: The common assumption is that the building has only planar roofs, based on which various model detection methods in pattern recognition can be adopted for building extraction.

The 3D reconstruction process from airborne LIDAR data is principally based on the segmentation of the raw data sets into building points; the segmentation refers to the task of dividing data into non-overlap homogeneous regions that constitute the complete data sets [7]. The efficiency and accuracy of the segmentation method results are one of the major challenges in 3D building reconstruction. Region growing methods and the clustering algorithm are the most used in literature for segmentation purpose. Region growing approaches (e.g., [15]) usually start with a selected seed point, calculate its properties, and compare them with adjacent points based on certain connectivity measures to form the region. Alternatively, the cluster techniques (e.g., [7,11]) first summarize the variability in the data by computing the attributes for all points and then group data that cluster together. Each point in the point cloud is classified into one of the clusters of predefined number based on its distances to the clusters' centroid. Other approaches, such as the Hough Transform [16] and random sample consensus (RANSAC) [17], can be used to extract straight lines from boundary points. Finally, the voxel-based algorithm (e.g., [18]) divides a point cloud into voxels with equal size, then the neighboring voxels with elevation differences of less than a threshold are classified iteratively into the same subset and segmented from other points.

This study aims to present an investigation and critical comparison of two different fully automatic approaches for roof segmentation used in 3D building reconstruction. In particular, we present a stable solution approach (a), described in Section 2.1, for building roof extraction based on a fuzzy c-means [19] that uses a potential-based clustering method for initial clusters center determination and clusters number determination [20,21]. At the end of the clustering processes, a density-based and connectivity analysis, as proposed by [7], is used to improve the results of the above clustering process through separation of the planar and coplanar planes. Moreover, a second approach (b), based on a region growing segmentation method [22] using smoothness constrain and curvature consistency refined with application of RANSAC [23] to remove any potential over-segmentation issues, is described in Section 2.2.

The roofs extracted by these two segmentation approaches were used for 3D building reconstruction. After the extraction of the boundary points, as described in [24], a 2.5D dual-contouring approach, proposed by [25], was adopted to create vertical walls connecting the extracted rooftops to the ground.

Both the proposed approaches have been evaluated (Section 3) in terms of geometry accuracy against the real measurements in two different case studies (in terms of types of urban development and ALS data input data resolution) in Matera (Italy) and Toronto (Canada). The results indicate that both approaches have precisely reconstructed the geometric features of the test building preserving topology. In particular, the approach (b) based on region growing segmentation has exhibited slightly better performance but required a computational time that is double that of the clustering-based approach (a). Finally, Section 4 presents the conclusion and main remarks on the investigation presented in this paper.

2. Materials and Methods

This section describes in detail the data used in this study (see Section 2.1) and the steps of the workflow pipeline used in the two proposed fully automatic segmentation approaches (see Sections 2.2 and 2.3) that process airborne LIDAR point cloud data for the purpose of building modeling and 3D reconstruction, as described in the Section 2.3.

2.1. LIDAR Data Set

The first data set was captured over downtown Toronto (Canada). Optech's ALTM-ORIONM was used to acquire the ALS data at a flying height of 650 m in six strips with a point density of about 6 points/m^2. The area contains a mixture of low- and high-storey (58) buildings, showing various degrees of complexity in rooftop structures. The reference for building detection and for 3D building reconstruction was generated by stereo plotting. The accuracy of well-defined points is 20 cm in planimetry and 15 cm in height. For more details, refer to [26] and the web site of the test [27]. The scene also contains trees and other urban objects.

The second data set was captured over a complex building in downtown Matera (Italy). The LIDAR survey in the historical center of Matera (Italy) (Figure 1a,b) was carried out by GEOCART S.p.A. using a full-waveform scanner [28], RIEGL LMS-Q560 on board a helicopter to obtain a higher spatial resolution. The flight was operated with a share of around 400 m, a speed of 25.7 m/s, and an opening angle at 60°. The scanner acquired data in the direction SN–EW, with a divergence of the radius 0.5 mrad, and a pulse repetition rate at 180,000 Hz. The average point density value of the dataset was about 30 points/m^2. The accuracy was 25 cm in x, y and 10 cm in z (altitude). The raw data of a small tile extracted by the survey have been orthorectified and radiometrically corrected in order to provide a ready-to-use point cloud to realize as output a group of watertight mesh models that could be used for various applications, such as energy demand estimation, classification of building types, visibility analysis, 3D cadastre, visualization for navigation, urban planning, emergency response, or flooding [4]. The LIDAR data is provided as a group of unorganized discrete points in which each individual point has x, y and z value, plus the intensity value that represents the reflective proprieties of surface encountering (Figure 1b).

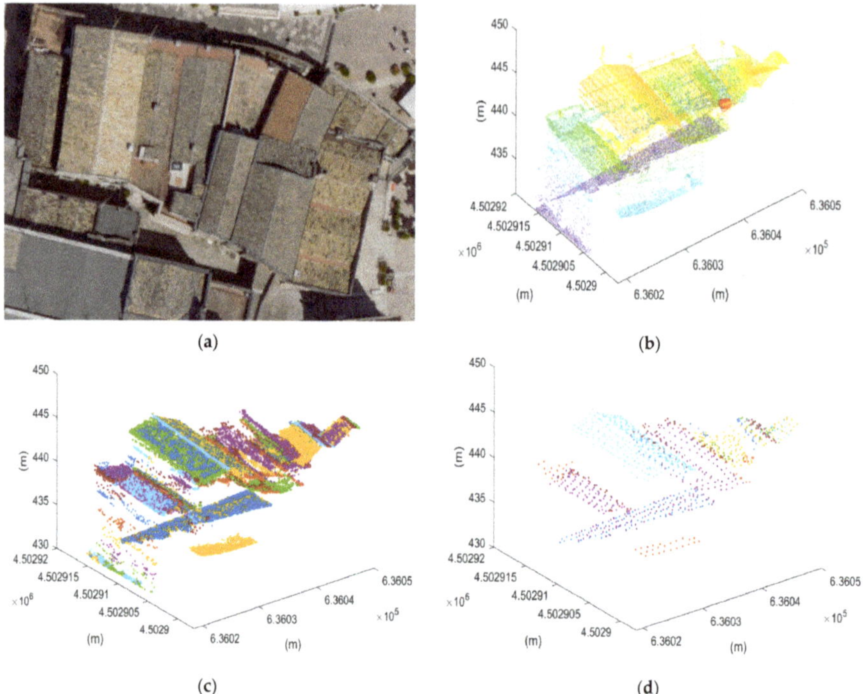

Figure 1. Clustering and segmentation of roof light detection and ranging (LIDAR) points. (**a**) orthopotho provided by GEOCART; (**b**) LIDAR points; (**c**) clustered points; (**d**) Planar segments.

2.2. Roof Segmentation Clustering Approach

In 3D building modeling, the segmentation process generally aims to find which LIDAR cloud points belong to which specific rooftop segments and to represent them with as many details as possible. In particular, the segmentation process begins with exploration of the proprieties of local distribution of the points' normal vector (N_i) that uniquely determines the direction of a roof plan in order to return the planes with the same normal vectors.

This section describes an approach in which this is treated as a cluster problem, as widely proposed in literature studies (e.g., [6,7,19,20]). In particular, the fuzzy c-means method was used to determine the clusters (see Figure 1c). This method belongs to the partitioning methods clustering category, relocating iteratively data points among various clusters until they achieve the greatest distinction. In this method, a data point does not belong to only one cluster entirely. Instead, it can belong to any of the clusters at a certain degree of belonging estimated measuring the similarity, that is, the inverse distance measure of each data point to the cluster centers.

The fuzzy c-means algorithm requires determination of the number of clusters and their approximate centers in order to start the iterative computation. In this context, the LIDAR data is pre-processed for c-means clustering using a potential-based clustering approach. The point with high potential (i.e., the highest number of data points within its (fixed-distance) sphere of influence) is used to determine the first cluster center; the potential (P_i^f) of data point (N_i) is calculated as:

$$P_i^f = \sum\nolimits_{i=1}^{j} e^{\{-\frac{4}{r_f^2}\|N_i - N_j\|^2\}},\qquad(1)$$

where j is the number of data points, N_j is the jth data point and r_f is the radius of the point N_i sphere of influence.

The other cluster center's potential (P_i^o) is then estimated based on the distance to the previously selected cluster centre(s) to reduce the possibility of two cluster centers being close:

$$P_i^o = P_i^f - P_f^* e^{\{-\frac{4}{r_0^2}\|N_i - N_f^*\|^2\}},$$
(2)

where N_f^* and P_f^* are the previously selected center and its potential, respectively. To avoid obtaining closely spaced cluster centers, we set r_0 to be somewhat greater than r_f, i.e., $r_0 = 1.5\, r_f$.

This process stops when the cumulative potential P_i^o reaches a threshold that is below 15% of the first potential P_i^f. If the potential falls between 15% and 50% of the first potential, we check if the data point provides a good trade-off between having a sufficient potential and being sufficiently far from existing cluster centers, as evaluated by [19].

This algorithm has been implemented iteratively, changing the value of the radius of the sphere of influence. This results in a set of scenarios depending on the radius r of the sphere of influence utilized, where the final number of the clusters is inversely proportional with the magnitude of the r. Selection of the best approximation value of the cluster numbers and the cluster center positions was determined by the likelihood (i.e., compactness) of each cluster, calculated as in the following equation:

$$d = \sum_{i=1}^{c} \frac{d^i}{c},$$
(3)

where d^i is the mean distance in cluster i of data points to its respective clusters and c the number of clusters.

Finally, rooftop segmentation was refined by the separation of parallel and coplanar planes as well, as proposed in [7], because the planes may have roof segments that are parallel to each other or roof segments that are mathematically the same but are separated spatially (see Figure 1d).

The flow of the proposed approach is depicted in Figure 2.

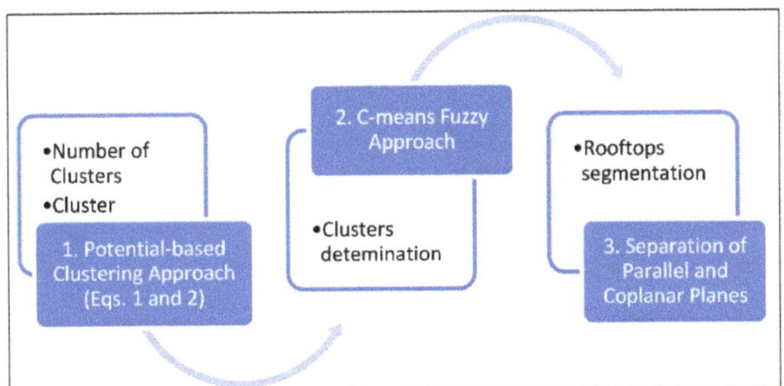

Figure 2. Schematic representation of the workflow of the clustering segmentation approach described in Section 2.2.

2.3. Roof Segmentation Region Growing Approach

The second approach uses the region growing segmentation method proposed by [24] in order to describe each individual building rooftop with the best spatial detail possible (see Figure 3). This region growing segmentation process uses the point normals (N_i) and their curvatures (C_i) to detect every significant feature on the rooftop:

$$C = \frac{\lambda_1}{\lambda_1 + \lambda_2 + \lambda_3}$$
(4)

where λ is the eigenvalue of the LIDAR points subdivided by their three dimensionalities.

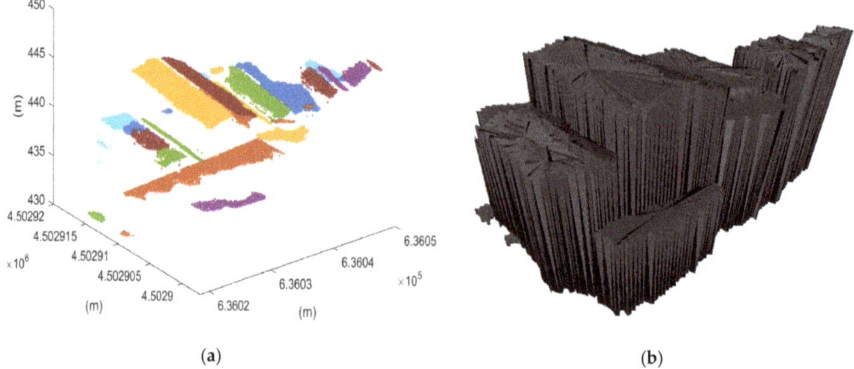

(a)   (b)

Figure 3. Region growing approach. (**a**) Segmentation of roof light detection and ranging (LIDAR) points displayed in Figure 1b. (**b**) 3D building reconstruction.

The process examines the points surface smoothness and picks the point with the smallest curvature value as a seed point. The algorithm then examines the local connectivity among the neighboring points, grouping the points with direction similar to the seed point normal, that is, lesser than a predetermined threshold (the angular difference threshold applied here is equal to 4°).

Among those points which have been grouped together by the seed point, points with curvature values lower than a predetermined threshold (equal to 0.01) are chosen as future seed points. The procedure continues in the same fashion and stops when all points have been visited. Finally, for each segmented region, RANSAC is applied to fit a virtual plane from the candidate points and then the points are forced to move on to this estimated plane in order to assign a perfect flatness property to each surface [24]. The main steps of the proposed approach are depicted in Figure 4.

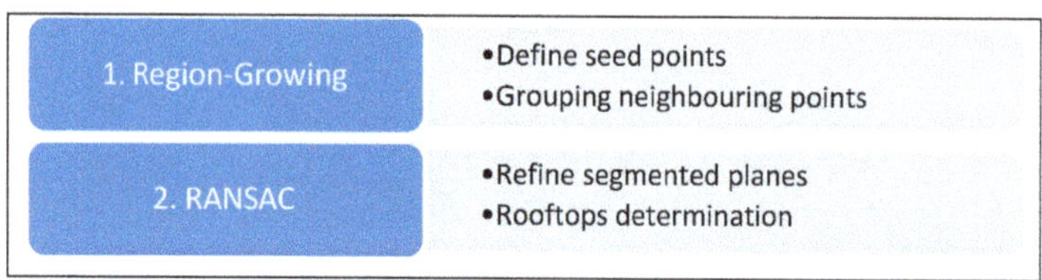

Figure 4. Schematic representation of the steps of the region-growing segmentation approach described in Section 2.3.

2.4. 3D Modeling

Successful extraction of major rooftop features, as proposed in Sections 2.2 and 2.3, is not sufficient to deliver 3D building modeling. Therefore, the next steps are (i) to produce possible boundary points for all features on the rooftops which are used, and (ii) to create vertical walls connecting rooftops to the ground.

Points on the boundaries of all detected parts on the rooftop are generated by applying rectilinear fitting: A 2D grid is overlaid on the LIDAR points in the x, y plane and each cell of the 2D grid is marked as being occupied and, thus, its boundary represents the shapes of all parts, if there are at least a minimum number of cloud points (based on their density). A robust 2.5D dual contouring method [25] is then utilized to generate facetized, watertight building models (see Figure 3b).

2.5. Evaluation of the Performance Measurements

The results of 3D building reconstruction of two different segmentation approaches described in Sections 2.2 and 2.3 were evaluated in terms of the geometrical accuracy of the roof polygons and/or final 3D building model. The mean, standard deviation, and Root Mean Square Error (RMSE) of the Euclidean distance (along x, y and z dimension) of each vertex (all the points) of the reconstructed 3D building model and the relative roof polygons and the nearest neighbors of the corresponding reference point were used:

$$RMSE = \sqrt{\sum_{i=1}^{n}(\hat{d}_i)^2}, \text{ where } \hat{d}_i = \sum_{i=1}^{n} \frac{\sqrt{[p_i(x) - r_i(x)]^2 + [p_i(y) - r_i(y)]^2 + [p_i(z) - r_i(z)]^2}}{n} \tag{5}$$

In Equation (5), $\hat{d}_i$ is the mean of the Euclidean distance (along x, y and z direction) between the i-th point p_i of each segmentation model and of the corresponding nearest neighbors point r_i of the reference data set.

3. Results

The two proposed fully automatic segmentation approaches were tested on two case studies with very different characteristics (e.g., different types of urban development and average point density value of the ALS input data) in order to increase the robustness and completeness of the proposed investigation.

First, an ALS data set with reference data made available via the International Society for Photogrammetry and Remote Sensing (ISPRS) web site [27] over downtown Toronto (Canada) was used to evaluate the performance of the two above-described segmentation approaches. This test case contained a mixture of 58 low and high-storey buildings; readers can refer to [26] for a more detailed description of characteristics and technical specifications of this benchmark dataset. Figure 5 shows a visual result of the 3D reconstruction using the proposed segmentation methods. Processing was conducted on a consumer laptop (Intel Core i7, 8G RAM). The average performances over the 58 buildings of the two roof segmentation approaches are shown in Table 1. The region growing segmentation approach exhibited slightly better performances but required a greater computational time than the clustering-based approach.

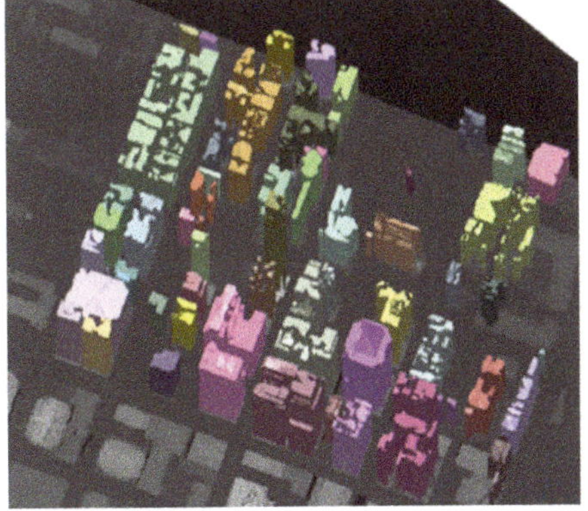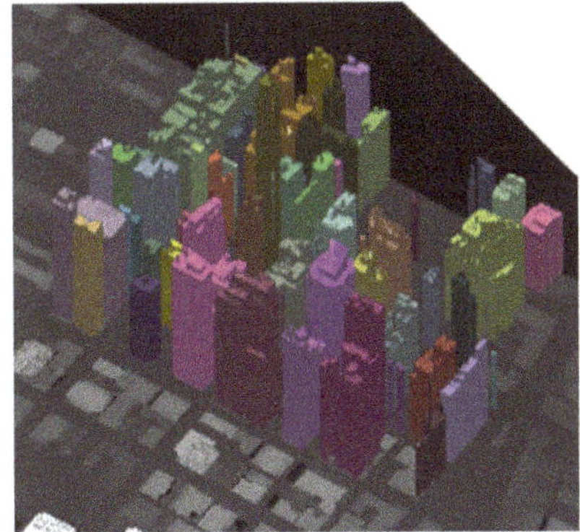

(a)

Figure 5. *Cont.*

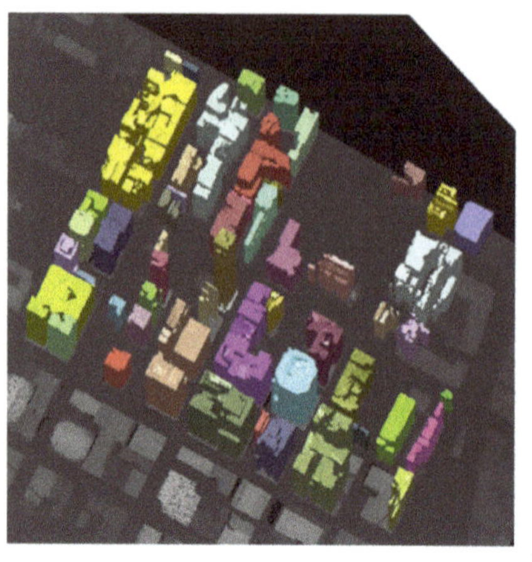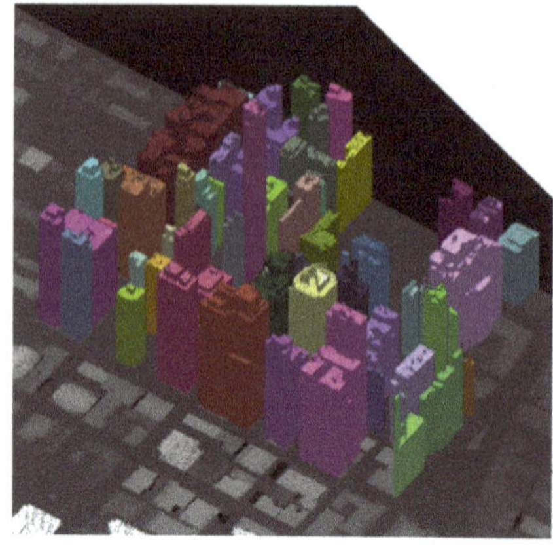

(b)

Figure 5. Results of 3d buildings reconstruction over downtown Toronto (Canada) using (**a**) the clustering roof segmentation approach and (**b**) the region growing approach.

Table 1. Average performances of the geometry accuracy on the 58 buildings over downtown Toronto (Canada) of each of the two segmentation approaches compared with the reference data.

Segmentation Approach	Average RMSE (m)	Average Dist. Mean (m)	Average Dist St. Dev. (m)	Average Computational Time (sec)
Clustering	3.81	3.20	2.08	17
Region-growing	2.65	2.09	1.61	26

A 3D reconstruction of one complex palace, constructed in the 1950s with a total surface of 3690 m^2, located in one of the main squares of the historical center of Matera (see Figure 6a), for which building a celerimetric survey made through a total station is available, was carried out from a LIDAR point cloud (see Figure 6a). This complex building was then manually reconstructed in 3D, as shown in Figure 6b, using commercial software that implements Building Information Modeling (BIM) technology [29]. This manually reconstructed 3D building is considered the gold standard because it is based on detailed survey measurements and, therefore, it can be compared with the outcomes of the two proposed segmentation approaches in order to evaluate their vertex geometry accuracy. Figure 6c,d shows the reconstruction results for the data collected for the building test case in Matera. Processing was conducted on a consumer laptop (Intel Core i7, 8G RAM) and is presented as solid models with simplified facades and faithfully reflected roof structure considering that the aims of these kinds of approaches are to realize an interactive visualization covering large areas. The performance of the two approaches, evaluated as described in Section 2.5, are shown in Table 2. The region growing segmentation approach exhibited slightly better performances but required a greater computational time (two times greater) than the clustering-based approach, similar to the performance in the previously described case study over downtown Toronto. Hence, the region growing approach analysing the LIDAR cloud points one by one can be more efficient to reach the best spatial detail possible but, at the same time, this process is more time-consuming. However, the potential-based method can also yield a stable estimate on the number of clusters and initial cluster centers which are needed for the following fuzzy k-means clustering calculation for an efficient segmentation process, saving computational resources. The better performances of both proposed methodologies on the Matera building with respect to the Toronto case study could be associated with the different quality of input ALS data adopted. This is confirmed by the fact that the minimum value of the RSME (using both segmentation methods) evaluated for each of the Toronto buildings is equal to 0.76 m, slightly higher than the value shown in Table 2 (i.e., 0.7 m).

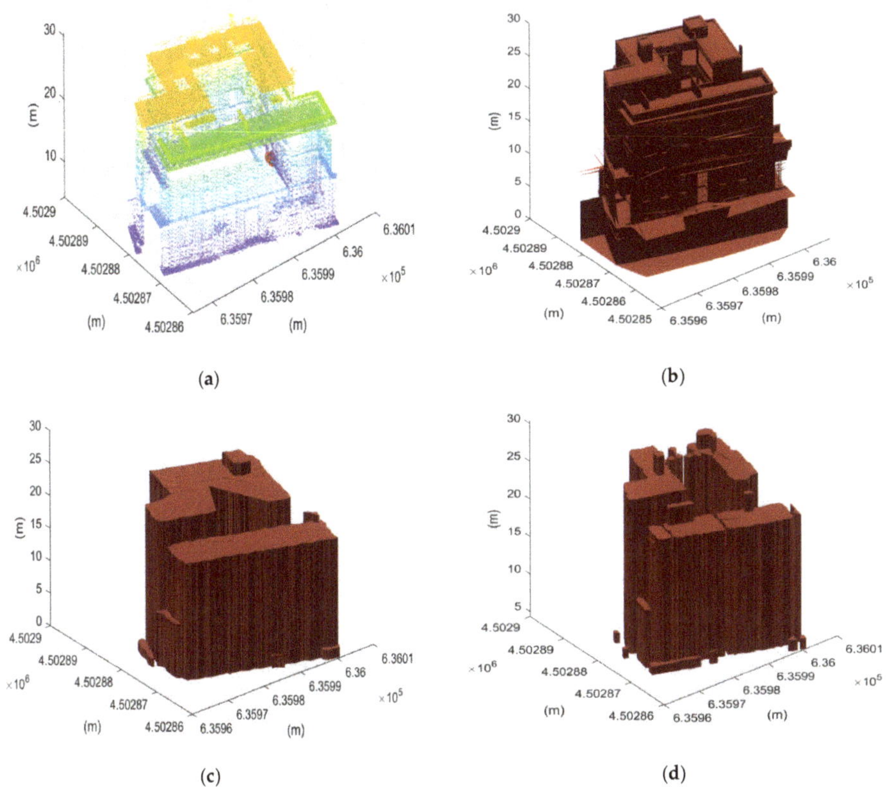

Figure 6. Results of the models for 3D reconstruction of the building located in Matera. (**a**) Light detection and ranging (LIDAR) points provided by GEOCART; (**b**) 3D Building Information Modeling (BIM) building reconstruction; (**c**) 3D building reconstruction using the clustering segmentation approach; (**d**) 3D building reconstruction using the region growing segmentation approach.

Table 2. Performances of the geometry accuracy of each of the two fully automatic reconstruction approaches compared with the 3D BIM model for the case study in Matera (Italy).

Segmentation Approach	RMSE (m)	Dist. Mean (m)	Dist St. Dev. (m)	Computational Time (sec)
Clustering	1.56	1.36	0.90	120
Region-growing	1.38	1.19	0.70	240

In both applications, each side of a rooftop is connected to the ground by a simple, vertical wall which is obviously not always indicative of the true architectural form. In addition, the 2.5D dual contouring method [15] is a robust algorithm although it does not respond to our ideal outline refinement.

4. Conclusions

Rottensteiner, F et al. [30] have tried to analyse a few of the great variety of detection and reconstruction applications from airborne laser scanning (ALS) proposed in the literature by identifying common problems of existing studies and by giving indications about the most promising applications. However, a research demand is still needed for comparing the results of different segmentation methodologies for 3D building reconstruction. Indeed, this study presents an investigation of fully automatic segmentation approaches for 3D building detection and modeling by processing airborne LIDAR point clouds. The first method proposed in this study for the extraction of rooftop patches uses a fuzzy c-means clustering method refined with the separation of planar and coplanar planes, which can be fairly easily accomplished based on planar equations and connectivity, respectively. In a second segmentation approach, a region growing based segmentation combined with RANSAC method was used to detect all significant features on the rooftop. Finally, the boundary regularization approach and the 2.5D dual-contouring method was adopted for the 3D modeling process using the outcome of each of these two segmentation approaches.

The results of both approaches were tested on two case studies that differ in their types of urban development and input data characteristics. The (i) benchmark LIDAR point clouds with the related reference data (generated by stereo plotting) over downtown Toronto (Canada) and (ii) the LIDAR data of a complex building in Matera (Italy) with the relative 3D BIM model (Building Information Modelling) (generated though celerimetric survey measurement with a total station) were used to evaluate the geometrical quality of roofs under different operating system of the above described segmentation approaches. Performances were evaluated in terms of computational time but also in terms of mean, standard deviation and Root Mean Square Error of the Euclidean distance (along x, y and z dimension) of each vertex (all the points) of the modeled roof polygons and the nearest neighbors of the corresponding reference point. The results of these two different case studies show that both methods reach good performance metrics in terms of geometry accuracy, demonstrating their transferability in other contexts. However, the approach based on region growing segmentation exhibited slightly better performances than the clustering-based approach and required greater computational time.

Funding: This research received no external funding.

Acknowledgments: The author gratefully acknowledges GEOCART S.p.A. for providing the LIDAR point clouds of Matera (Italy) and the author would like to acknowledge the provision of the Downtown Toronto data set by Optech Inc., First Base Solutions Inc., GeoICT Lab at York University, and ISPRS WG III/4.

Conflicts of Interest: The author declares no conflict of interest.

References

1. Bonczak, B.; Kontokosta, C.E. Large-scale parameterization of 3D building morphology in complex urban landscapes using aerial LiDAR and city administrative data. *Comput. Environ. Urban Syst.* **2019**, *73*, 126–142. [CrossRef]
2. Ma, Z.; Liu, S. A review of 3D reconstruction techniques in civil engineering and their applications. *Adv. Eng. Inf.* **2018**, *37*, 163–174. [CrossRef]
3. Wang, R. 3D building modeling using images and LiDAR: A review. *Int. J. Image Data Fusion* **2013**, *4*, 273–292. [CrossRef]
4. Zlatanova, S.; Çöltekin, A.; Ledoux, H.; Stoter, J.; Biljecki, F. Applications of 3D city models: State of the art review. *ISPRS Int. J. Geo Inf.* **2015**, *4*, 2842–2889.
5. Wang, R.; Peethambaran, J.; Chen, D. LiDAR point clouds to 3-D urban models: A review. *IEEE J. Sel. Top. Appl. Earth Obs. Remote Sens.* **2018**, *11*, 606–627. [CrossRef]
6. Shan, J.; Sampath, A. Building extraction from LiDAR point clouds based on clustering techniques. *Topogr. Laser Ranging Scanning* **2010**, *11*, 421–444.
7. Sampath, A.; Shan, J. Segmentation and reconstruction of polyhedral building roofs from aerial lidar point clouds. *IEEE Trans. Geosci. Remote Sens.* **2010**, *48*, 1554–1567. [CrossRef]
8. Pahlavani, P.; Amini Amirkolaee, H.; Bigdeli, B. 3D reconstruction of buildings from LiDAR data considering various types of roof structures. *Int. J. Remote Sens.* **2017**, *38*, 1451–1482. [CrossRef]
9. Jung, J.; Jwa, Y.; Sohn, G. Implicit regularization for reconstructing 3D building rooftop models using airborne LiDAR data. *Sensors* **2017**, *17*, 621. [CrossRef]

10. Cao, R.; Zhang, Y.; Liu, X.; Zhao, Z. Roof plane extraction from airborne lidar point clouds. *Int. J. Remote Sens.* **2017**, *38*, 3684–3703. [CrossRef]

11. Filin, S.; Pfeifer, N. Segmentation of airborne laser scanning data using a slope adaptive neighborhood. *ISPRS J. Photogramm. Remote Sens.* **2006**, *60*, 71–80. [CrossRef]

12. Haala, N.; Brenner, C. Extraction of buildings and trees in urban environments. *ISPRS J. Photogramm. Remote Sens.* **1999**, *54*, 130–137. [CrossRef]

13. Fischer, A.; Kolbe, T.H.; Lang, F.; Cremers, A.B.; Förstner, W.; Plümer, L.; Steinhage, V. Extracting buildings from aerial images using hierarchical aggregation in 2D and 3D. *Comput. Vis. Image Underst.* **1998**, *72*, 185–203. [CrossRef]

14. Suveg, I.; Vosselman, G. 3D reconstruction of building models. *Int. Arch. Photogramm. Remote Sens.* **2000**, *33*, 538–545.

15. Cheng, L.; Wu, Y.; Wang, Y.; Zhong, L.; Chen, Y.; Li, M. Three-dimensional reconstruction of large multilayer interchange bridge using airborne LiDAR data. *IEEE J. Sel. Top. Appl. Earth Obs. Remote Sens.* **2015**, *8*, 691–708. [CrossRef]

16. Morgan, M.; Habib, A. Interpolation of lidar data and automatic building extraction. *ACSM-ASPRS Annu. Conf. Proc.* **2002**.

17. Fischler, M.A.; Bolles, R.C. Random sample paradigm for model consensus: A apphcatlons to image fitting with analysis and automated cartography. *Commun. ACM* **1981**, *24*, 381–395. [CrossRef]

18. Dimitrov, A.; Golparvar-Fard, M. Segmentation of building point cloud models including detailed architectural/structural features and MEP systems. *Autom. Constr.* **2015**, *51*, 32–45. [CrossRef]

19. Jain, A.K.; Murty, A.M.N.; Flynn, P.J. Data clustering: A review. *ACM Comput. Surv.* **1999**, *31*, 264–323. [CrossRef]

20. Chiu, S. Fuzzy model identification based on cluster estimation. *J. Intell. Fuzzy Syst.* **1994**, *2*, 267–278.

21. Tibshirani, R.; Walther, G.; Hastie, T. Estimating the number of clusters in a data set via the gap statistic. *J. R. Stat. Soc. Ser. B Stat. Methodol.* **2001**, *63*, 411–423. [CrossRef]

22. Rabbani, T.; van Den Heuvel, F.; Vosselmann, G. Segmentation of point clouds using smoothness constraint. *Int. Arch. Photogramm. Remote Sens. Spat. Inf. Sci.* **2006**, *36*, 248–253.

23. Zuliani, M. Ransac forDummies, with Examples Using the RANSAC Toolbox for Matlab and More. 2014. Available online: http://citeseerx.ist.psu.edu/viewdoc/download?doi=10.1.1.475.1243&rep=rep1&type=pdf (accessed on 1 September 2018).

24. Sun, S.; Salvaggio, C. Airborne LiDAR point clouds. *IEEE J. Sel. Top. Appl. Earth Obs. Remote Sens.* **2013**, *6*, 1440–1449. [CrossRef]

25. Zhou, Q.-Y.; Neumann, U. 2.5D dual contouring: A robust approach to creating building models from aerial lidar point clouds. *Comput. Vis. ECCV* **2010**, *6313*, 115–128.

26. Rottensteiner, G.; Sohn, M.; Gerke, J.D.; Wegner, U.; Breitkopf, J. Jung results of the ISPRS benchmark on urban object detection and 3D building reconstruction. *ISPRS J. Photogram. Rem. Sens.* **2014**, *93*, 256–271. [CrossRef]

27. ISPRS. Web Site of the ISPRS Test Project on Urban Classification and 3D Building Reconstruction. 2013. Available online: http://www2.isprs.org/commissions/comm3/wg4/tests.html (accessed on 15 October 2019).

28. Mallet, C.; Bretar, F.; Roux, M.; Soergel, U.; Heipke, C. Relevance assessment of full-waveform lidar data for urban area classification. *ISPRS J. Photogramm. Remote Sens.* **2011**, *66*, S71–S84. [CrossRef]

29. Eastman, C.; Teicholz, P.; Sacks, P.; Liston, K. *BIM Handbook: A Guide to Building Information Modeling for Owners, Managers, Designers, Engineers and Contractors*; Wiley: Hoboken, NJ, USA, 2008.

30. Rottensteiner, F.; Sohn, G.; Gerke, M.; Wegner, J.D. Theme section "Urban object detection and 3D building reconstruction". *ISPRS J. Photogramm. Remote Sens.* **2014**, *93*, 143–144. [CrossRef]

Review

Status of BIPV and BAPV System for Less Energy-Hungry Building in India—A Review

Pranavamshu Reddy [1], M. V. N. Surendra Gupta [2], Srijita Nundy [3], A. Karthick [4] and Aritra Ghosh [5,*]

1   Electronics and Communication Engineering, SASTRA University, Tirumalaisamudram Thanjavur 613401, Tamil Nadu, India; pranavamshu4@gmail.com
2   Academy of Scientific and Innovative Research, CSIR-SERC, Chennai 600113, Tamil Nadu, India; mvnsgupta4805@gmail.com
3   School of Advanced Materials Science and Engineering, Sungkyunkwan University, Suwon 16419, Korea; srijita.31121990@gmail.com
4   Department of Electrical and Electronics Engineering, KPR Institute of Engineering and Technology, Avinashi Road, Arasur, Coimbatore 641 407, Tamilnadu, India; karthick.power@gmail.com
5   Environment and Sustainability Institute (ESI), University of Exeter, Penryn Campus, Penryn TR10 9FE, UK
*   Correspondence: a.ghosh@exeter.ac.uk

Received: 25 February 2020; Accepted: 24 March 2020; Published: 29 March 2020

Abstract: The photovoltaic (PV) system is one of the most promising technologies that generate benevolent electricity. Therefore, fossil fuel-generated electric power plants, that emit an enormous amount of greenhouse gases, can be replaced by the PV power plant. However, due to its lower efficiency than a traditional power plant, and to generate equal amount of power, a large land area is required for the PV power plant. Also, transmission and distribution losses are intricate issues for PV power plants. Therefore, the inclusion of PV into a building is one of the holistic approaches which reduce the necessity for such large land areas. Building-integrated and building attached/applied are the two types where PV can be included in the building. Building applied/attached PV(BAPV) indicates that the PV system is added/attached or applied to a building, whereas, building integrated PV (BIPV) illustrates the concept of replacing the traditional building envelop, such as window, wall, roof by PV. In India, applying PV on a building is growing due to India's solar mission target for 2022. In 2015, through Jawaharlal Nehru National Solar Mission, India targeted to achieve 100 GW PV power of which 40 GW will be acquired from roof-integrated PV by 2022. By the end of December 2019, India achieved 33.7 GW total installed PV power. Also, green/zero energy/and sustainable buildings are gaining significance in India due to rapid urbanization. However, BIPV system is rarely used in India which is likely due to a lack of government support and public awareness. This work reviewed the status of BIPV/BAPV system in India. The BIPV window system can probably be the suitable BIPV product for Indian context to reduce the building's HVAC load.

Keywords: India; BIPV/BAPV; BIPV-Glazing; JNNSM; MNRE; Zero energy-building

---

1. Introduction

India's energy consumption has increased to 931 billion kWh which is double than of the level marked in the year 1990 making it one of the largest energy consumers in the world along with China, the USA and, Russia. In 2014, India's per capita electricity consumption was 900 kWh which was 1/3 of the average worldwide consumption [1,2]. Also, compared to 1971, Indian per capita energy use has increased from 3116.84 to 7408.31 kWh in 2014. India's energy sources primarily depend on non-renewable coal-based sources which contribute to a massive amount of greenhouse gases [3–5].

Currently, India is facing urbanization due to the migration of people into larger cities from smaller towns and villages. This transition enhances the necessity of developing new buildings. During 2014-2015, India consumed almost 840 million m^2 floor space for commercial use. Buildings in India, consume 29% of the total energy, of which residential contributes to 20% and commercial to 9% [6]. In residential buildings, lighting and space cooling accounts to one-third of the energy consumption (1–3 kWh/m^2/month), whereas commercial buildings consume two-third of the total energy (5–25 kWh/m^2/month) [7]. The phenomenal growth in the building sector will be witnessed by the year 2030 with an annual building rate of 700–900 million sq. m. These buildings consume a considerable amount of energy for heating, ventilation and air conditioning (HVAC) load demand. Further, the indoor air condition rate is growing at a rate of 30% every year. Projected energy usage for 2050, with the current scenario, shows an 85% increment compared to the energy level in 2005 [8].

The present energy consumption scenario, along with future projections, has forced India to take some necessary actions. In 2015, the international energy association (IEA) has set a target to limit the ambient temperature increment to below 2 °C than the pre-industrial levels. Hence, India should focus on finding out the energy-efficient ways to generate power and also reduce building energy consumption rates. Fortunately, India is blessed with high solar radiation, which receives 6 billion GWh equivalent energy potential per year. The average incident solar radiation in India is 5.1 kWh/m^2/day (with large regional differences). This makes India deploy solar photovoltaic (PV) technology to meet the IEA target. The PV device is one of the most promising renewable energy technologies, which converts solar energy into environment-friendly electrical energy by using abundant incident solar radiation. Replacing fossil fuel-generated power by secure, clean and suitable PV generated power can mitigate issues like climatic changes [9,10].

The Ministry of Power, which controls the power sector in India, created an impressive mission through Jawaharlal Nehru National Solar Mission (JNNSM). Previously, JNNSM has set a target to install a PV capacity of 22 GW by the year 2022 which later increased to a more ambitious target of 100 GW [11]. Subsequently, to reduce building energy consumption and generate power from renewable sources in the buildings; zero energy buildings (ZEB) or net-zero energy buildings are also getting a promotion. Hence, addition of the PV system into the building is one of the most holistic approaches, where, PV will generate a benevolent amount of energy, sufficient for the building-energy requirements. The inclusion of PV technologies into buildings include building-integrated photovoltaics (BIPV) and building-applied photovoltaics (BAPV). For the BIPV system, the PV system replaces the traditional building envelopes, such as windows, roofs, walls and itself acts as a building envelope, whereas for the BAPV system, PVs are applied or attached to the building walls or roofs. Both BAPV and BIPV works as an onsite green power generation, reducing the transmission losses, and improving the building's overall performance.

In this paper, various technologies involving BIPV and BAPV approaches have been discussed and their potential application for Indian context has been critically analyzed in detail. Moreover, solar potential and PV power electricity market in India are also discussed.

2. PV Technologies for BIPV/BAPV

Presently PV technologies include first-generation opaque silicon type, second-generation transparent or semitransparent thin film and third or emerging types [12]. Until now, the first generations are employed for BIPV and BAPV applications, whereas second and third generations are primarily considered for BIPV application.

Crystalline silicon (c-Si) PV cells are the most widely used and predominant technology in the market due to their mature and long-term durability. Monocrystalline PV cells are made from a single crystal, developed using the Czochralski process with the best-reported efficiency of nearly 22%. Polycrystalline solar cells are developed by melting several fragments of silicon together to form a wafer. Typical efficiency is in the range of 14–18% for polycrystalline PV cells, which is less efficient than the monocrystalline counterparts, since electrons have less freedom of movement due to grain boundaries of many crystals in each cell. However various anti-reflective coatings can be applied onto the surface to change the color of the PV cells. Presently colored silicon PV is also under investigation [13,14]. The major constraints of crystalline silicon PV cells are power losses due to the shading and at elevated temperature [15–22].

Thin films include (i) amorphous silicon (a-Si) (ii) Copper – Indium Selenide (CIS) or Copper-Indium-Gallium- Selenide (CIGS), (iii) Cadmium-Telluride (CdTe). The thickness of the film could be a few nanometers to micrometers. These technologies have meager efficiencies in comparison to c-Si, typically 11–12%. However, they have several advantages such as (a) less loss in performance under overcast cloudy climatic conditions and partial shading from obstacles [23,24] (b) employ lower semiconductor material and hence lower production cost (c) manufacture of transparent or translucent modules using laser scribing [25–27]. Amorphous silicon is the non-crystalline form of silicon, with atoms disoriented in a random network structure. The major advantage of it is being able to be deposited as thin films on to a moldable substrate like plastic at less than 300 °C of manufacturing temperature. Moreover, its absorptivity is higher (~40 times) and needs only 1% (about 1 μm) of material of crystalline silicon, which results in lower making cost/unit-area. Due to its flexible nature, it can be molded into any suitable complex shape for building integration. Although it has high efficiency in comparison to other thin-film technologies, it suffers from degradation due to hydrogenation (Staebler-Wronski effect) [28–33]. Cadmium telluride (CdTe) is a single-junction solar cell having 1.45 eV bandgap energy. It is a direct bandgap semiconductor nearly ideal for optimal conversion of solar radiation into electricity. An efficiency exceeding 20% has been reported CdTe PV. The major limitations of CdTe cells are its instability and toxicity of cadmium which makes it less suitable for PV application. Copper Indium Gallium Diselenide (CIS) is a polycrystalline compound consisting of copper, indium, gallium, sulphur and selenide elements, with the highest reported conversion efficiency of about 25% in combination with perovskites [34] CIS has high light absorptivity and 0.5 μm of CIS can absorb 90% of the solar spectrum [35]. Similar to other thin-film technology, CIGSs are semi-transparent and flexible.

Emerging third generations are gaining importance due to their low fabrication cost, and transparent and semitransparent makes them a potential candidate for aesthetic building integration. Organic photovoltaics uses organic polymer as the light-absorbing layer. Organic PVs are lightweight, and flexible which allows them to be applied in building as a BIPV system [36–38]. O'Regan and Gratzel carried out seminal work on dye-sensitized solar cells (DSSC) [39]. Since its inception, extensive research was carried out to improve the efficiency and stability of DSSC. DSSCs are considered for BIPV application due to its simpler and low-cost fabrication process, flexible, have potential to operate at diffuse solar radiation [40,41]. Colored and semi-transparent windows are popular for BIPV application [42,43]. However, factors inhibiting to its commercialization are long term- stability and durability. Table 1 listed the advantage and disadvantages of various PV technologies. Recently, Perovskite PV gained attention due to its efficiency improvement in 10 years. However, they are mostly operated and fabricated at inert atmospheric condition. Tunable transparency [44], and low temperature fabrication [45] makes it fascinating to researcher for BIPV application [46,47].

Table 1. Advantages of disadvantages of various solar cells.

Solar cell	Advantages	Disadvantages
Monocrystalline silicon solar cells [17,48,49]	<ul><li>Matured PV technologies.</li><li>Highly durable.</li><li>Suitable for BAPV application.</li></ul>	<ul><li>There is a lot of waste material when silicon is removed during processing.</li><li>At high temperature, performance degrades.</li><li>Opaque in nature, hence less suitable for artistic BIPV application.</li></ul>

Table 1. *Cont.*

Solar cell	Advantages	Disadvantages
Polycrystalline silicon solar cells [50–52]	• The production process is simpler than the monocrystalline cells. • Highly durable.	• Due to low level of silicon purity, performance is only around 13–16%.
Amorphous silicon solar cells [53–55]	• Low manufacturing costs. • The cell can be produced in various shapes. • At high temperature, performance degradation is lower than crystalline silicon.	• The efficiency is typically 6–8%. • They have shorter lifetime compared to other solar cells. • Required twice, the space to get same PV power than that of crystalline silicon.
CdTe solar cells [56–61]	• Cadmium is abundant. • The manufacturing process is simple. • It can absorb light of shorter wavelength.	• The efficiency operates in the range 9–11%. • Tellurium is not abundant. • Cadmium telluride is toxic and not environmentally benign.
CIGS solar cells [35,52,62–67]	• CIGS solar cells use lower levels of cadmium, in the form of cadmium sulphide. • CIGS solar cell substrates are more versatile in comparison with c-Si. • CIGS solar panels show better resistance to heat compared to silicon solar cells.	• The efficiency ranges in between 12–14%. • High fabrication and production costs.
Organic solar cells [36–38]	• The PV modules are low weight and flexible. • Lower production costs than traditional inorganic technologies, such as silicon solar cells.	• Lifetime is short. • Very low efficiency around 4–5%.
DSSC [68–70]	• It consists of low-cost materials easy fabrication. • It works even in low light conditions such as the cloudy weather. • Tunable transparency is possible by tuning the thickness and dye. • Use of flexible substrate makes it flexible and suitable for BIPV.	• Low efficiency around 12%. • The liquid electrolyte has temperature stability problems. • The liquid will quickly dry up. • Long term stability is questionable.

Table 1. *Cont.*

Solar cell	Advantages	Disadvantages
Perovskite solar cells [71–73]	• Perovskite uses smaller quantity of material to absorb the equivalent amount of light in comparison to c-Si. • Perovskite materials, like methylammonium lead halides are cheap and easy to produce. • Semi-transparent/transparent cells are possible, which makes it suitable for aesthetic building application.	• Not stable at ambient condition. • Not fully matured technology. • Thermal performance of this technology is not well.

3. Building Integrated and Applied Photovoltaic (BIPV/BAPV) Technologies

BAPVs are an addition to the traditional or new PV system, on an existing or new building whereas BIPVs replace the existing traditional building envelope, such as window, roof, and wall [74]. Hence, BIPV has a greater impact on the building's indoor environment. BIPVs are often transparent or semi-transparent by nature, which allows incident daylight and solar heat to pass through, thereby directly modifying the indoor ambiances. Additionally, it also has a variety of capabilities like control solar heat gain or loss, daylight glare and offset the window, roof or wall material cost [75]. On the other hand, BAPVs have no such contribution to the building environment other than the production of green power. Currently, available BIPV products include BIPV tile, foil and glazing [76,77]. BIPV foils and tiles are primarily applied on the roof while BIPV glazings are mostly employed for vertical semi-transparent and transparent windows, façade and wall applications. Presently, 80% of the BIPV market contributes to rooftop-mounted and only 20% of it is in accord for façade-mounted [2,78–82]. Generally, rooftops, standing without any hindrance from nearby tall buildings or trees, are the ideal solutions to harvest the best energy when pitched at certain elevation angles. BIPV foil products are best suited for building applications due to their flexibility and light-weight properties. PV cells for BIPV products are mostly thin films, which possess low power generation due to the high electrical resistance of the thin film. However, due to a low temperature coefficient of thin-film BIPV foil, power degradation is comparatively less at high temperatures to silicon types. Alwitra GmbH and Co, which uses amorphous silicon cells and Uni-Solar cells are the present manufacturers of BIPV foil. Next, BIPV tiles are most prominently used as roof integration, which includes covering the entire roof or selected part of the roof with BIPV tiles. Some of these tiles also appear to be similar to that of a ceramic curved tile, which might be aesthetically pleasing but are not effective in terms of power generation, due to its curved surface area [76,77]. SRS Energy, Solar Century, Luma solar, Suntegra and, Sunflare Tesla are few of the present BIPV tile manufacturers.

Figure 1 shows the presently available BIPV and BAPV products. BIPV windows are one of the most fascinating applications, responsible for maintaining visual comfort between the external world and building interior, modulating available daylight and heat. Crystalline silicon [83,84], amorphous silicon [85–88], CdTe [89,90], DSSC [91,92] and perovskite [44] are the few materials, which when intensely investigated for BIPV window applications (shown in Figure 2), have found impressive possibilities towards building integration.

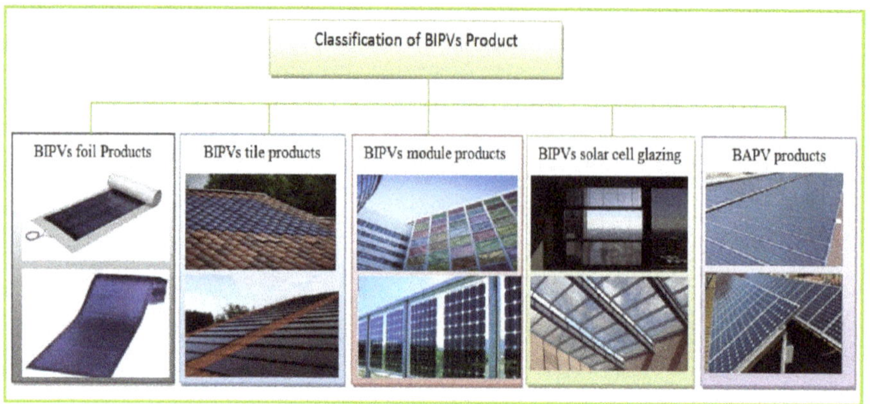

Figure 1. Major BIPV and BAPV products [79].

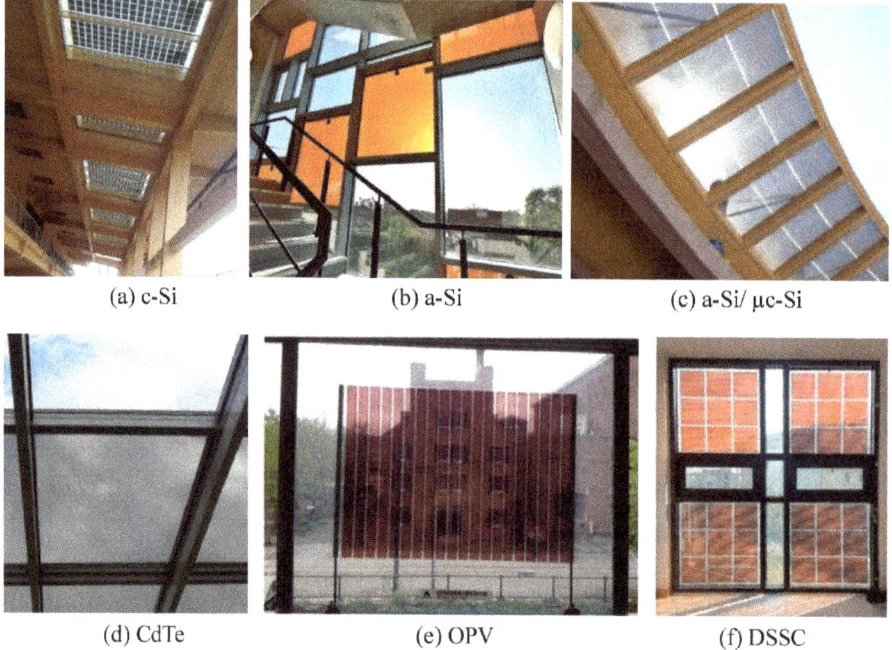

Figure 2. Window integrated with different types of PV cell materials [93].

Concentrator-based BIPV windows are also very attractive for less energy-hungry building integration. Low concentrating compound parabolic concentrator (CPC) [94] and luminescent solar concentrators (LSC) [95] are now dominating the major building-integrated concentrating photovoltaic research activity. Low concentrators are static which reduces the cost of the expensive solar tracker. The thermal effect is lower than a high concentrator, due to a low concentrator on PV cells, which reduces the necessity of a cooling system and makes a low concentrator a suitable candidate for building's window and façade application [96–98]. For northern latitude location, diffuse solar radiations are higher, CPC and LSC both work efficiently. Concentrating PV came into a scenario to reduce the usage of the costly silicon material by replacing low-cost material, which concentrates a higher amount of incident solar light on a lesser PV material [99]. For crystalline silicon-CPC-based BIPV window, regular distribution of spacing between PV cells offer semi-transparency (silicon solar cells are opaque), as shown in Figure 3. Different geometries of CPCs were investigated for BIPV window and façade application [100–103]. Recently, the performance of DSSC and perovskite solar cells was also investigated, using low concentrating CPC, which enhanced the PV performance than non-concentrating counterpart [104,105].

(a)   (b)

Figure 3. (**a**) CPC based BIPV, (**b**) Semi-transparent building blocks using CPC-silicon PV (image courtesy Build Solar).

A typical LSC consists of a glass or plastic-based square/rectangular-shaped waveguide luminophores, which absorbs a short-wavelength photon and convert them into long-wavelength. Further, due to total internal reflection, these photons finally reach the PV cells attached at the edge of the waveguide [106–110], as shown in Figure 4. The advantage of LSC-BIPV system is that the PV cells are placed at the edge, which does not create an obstacle for viewing. Also, this waveguide plate can be made semi-transparent to fully-transparent or different colors, which is aesthetic for building application and suitable for building window integration [111–114]. Promising results, using LSC-thin film integration, was also reported by [115]. LSC does not possess any thermal effect on PV cells, which is an added advantage over low concentrating CPC.

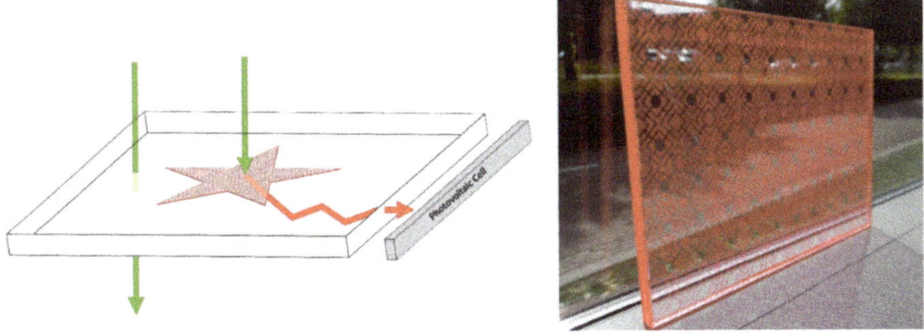

Figure 4. Working principle of inkjet-printed luminescent solar concentrator and photograph of a printed A4 sized luminescent solar concentrator [109].

The performance degradation of both BIPV and BAPV is possible at higher ambient temperature and exposure to higher incident radiation, which increases the PV cell's temperature. Crystalline silicon and thin-film both work, with poor efficiency, at higher PV cell temperature. Thermal regulation BAPV system is possible by employing forced water flow, forced airflow or phase change material (PCM) at the back of the system [116,117]. At the back of the BPAV system, copper pipes are integrated to flow the air [118–120] or water [121,122]. This typical BAPV is also known as BAPV-thermal (BAPV/T) water or air collector, where water or air will extract the additional heat energy from the PV system and allow PV system to operate efficiently. Hot air or water can be used by building purposes [123]. Notably, researchers often misuse the BIPV term [121,122,124–130]. As BIPV is attached as a building envelope, natural airflow and PCM are the only two available and investigated options to diminish the elevated PV cell temperature [131–133]. The inclusion of collector to BIPV was forced to compromise with buildings aesthetic. For a system which is BAPV/T system is most often referred to as BIPV/T in the articles. Hence, care should be taken when referring to a PV/T system as BAPV/T or BIPV/T. A typical BAPV/T with a water flow system is shown in Figure 5. Details of BIPV and BAPV are listed in Table 2.

Figure 5. BAPV/T system installed at Sodha BERS complex, Varanasi (25.33° N, 82.99° E) [134].

Table 2. Details of different BIPV and BAPV products.

Product	Type of Cell	Particular Purpose
BIPV window	1st generation 2nd generation 3rd generation	Control daylight, solar heat Aesthetic application Power generation gets lower priority
BIPV foil/tiles	2nd generation (mostly available) 3rd generation (suitable but stability should be improved	Works as building shading from the harsh external environment Power generation gets lower priority
Spaced type concentrating BIPV	1st generation (experimentally validated/commercial product is available) 2nd generation (no report) 3rd generation (experimentally explored in the lab)	Improve the power generation Reduce the cost of the system Spaced type allow daylight suitable for passive house application/zero energy application
BAPV/T	1st generation 2nd generation	Suitable for rooftop application

4. Potential of BIPV/BAPV in India

India lies between 68°7′ to 97°25′ east longitude and 8°4′ to 37°6′ north latitude, has 2.9 million Km^2 of landmass, and is the seventh largest country in the world. It is in the tropical region and receives maximum solar radiation in summer, and experiences about 300 sunny clear days in a year. Ambient conditions vary from 45 °C in summer while 4 °C in winter and has a hot-dry, warm-humid, composite, temperate, and cold climatic zones [135]. India's rich solar radiation profile shows 4.5–5.0 KWh/m^2/Day of annual average direct normal irradiance in most of the Indian states and around 5.0–5.5 KWh/m^2/Day average global horizontal irradiance [136,137]. This makes India one of the most potential candidates to contribute to PV power generation. Figure 6 shows the solar radiation intensity throughout India. India's projected electricity demand in 2047 is expected to be 5518 TWh. India's present energy demand is supplied by 70% of imported crude oil and coal. Indian thermal power plants, that are run by coal, are the most inefficient ones. Hence, in order to become an energy-secured country and dependent from oil-import, India should use its solar PV power potential. Although, India has a higher solar radiation, being a developing country, PV power generation faces issues, including high initial installation cost of PV power generation, the lack of suitable storage devices or unavailable during an instant power supply demand. Backup power supply from fossil fuel-generated kerosene oil lamps, diesel generators, etc. are still low cost in India [138]. However, this fossil fuel generated power emits considerable amount of green house gas (GHG). India is committed to lowering its GHG to 30–33% by 2030, compared to 2005 level.

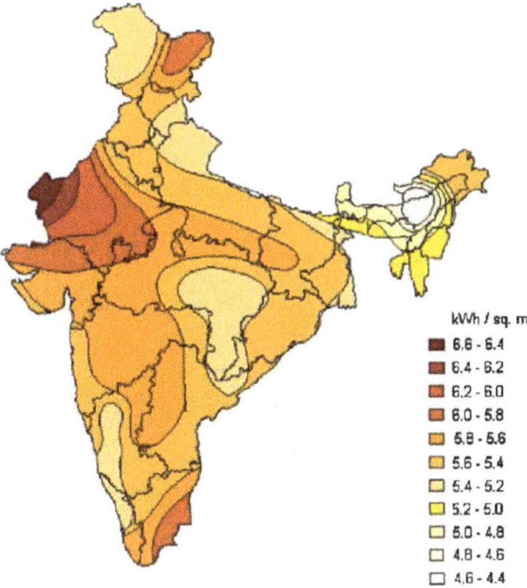

Figure 6. Physical map Indian solar radiation [139].

Interestingly, energy generation from PV devices was in discussion in India since 1960. However, progress was limited until 2010 [140,141]. The first significant move was taken in 2010 to priorities the PV power generation through the National Action Plan on Climate Change, by launching the JNSSM scheme [142]. In 2010, the total installed capacity from PV was only 39.6 MW. After Prime Minister Narendra Modi came into power, India's 2022 target changed from 20 GW to 100 GW, which included grid-connected projects, off-grid projects, and solar parks. It was also fixed that out of 100 GW, rooftop PV should produce 40 GW by 2022 [143]. Under the JNSSM mission, grid-connected rooftops and small solar power plant programs have been launched to obtain the 40 GW rooftop power generation. The minimum and maximum limit of installing PV power capacity are to 1 kWp, and 500 kWp, respectively [139,144]. India has 29 states and 7 union territory out of which Madhya Pradesh, Gujarat Ladakh, Andhra Pradesh, Maharashtra, and Rajasthan receive the maximum amount of average annual solar radiation, as compared to other states of India. Gujarat is the first Indian state which implemented a solar policy in the year 2009, well before the initiation of JNNSM. Rajasthan started its solar mission in 2011 to meet the national target. Karnataka started its solar mission for the period of 2014 to 2021. Madhya-Pradesh started its solar policy in 2012 and provided the incentives and benefits to the Private Sector to encourage the PV installation [145]. To initiate the government target, several commercial investors came in front. Cleanmax solar had invested Rs 600 crores to set up a 150 MW solar farm in Sirsa District, Haryana (near to New Delhi, 29.05° N, 76.08° E) on a stretch of 600 acres of land. Bharathi Cement had commissioned a 10 MW solar power plant in the manufacturing facility, located at Kadapa in Andhra Pradesh (14.46° N, 78.82° E). The plant is expected to generate 1.6 crore units of electric power annually, and help to reduce Bharathi's overall energy costs by reducing its dependence on thermal power. Maruti Suzuki India had planned to invest Rs 24 crore ($3366k) to set up a 5 MW solar power plant at its Gurugram (28.45° N, 77.02° E) facility. The plant would help to lower CO_2 emissions by 5,390 tonnes annually in 25 years. ReNew Power had commissioned 300 MW solar plant at Pavagada Solar Park in Tumkur district in Karnataka (13.37° N, 76.64° E). The solar power plant could reduce 0.6 million tonnes of CO_2 emission per year. The plant uses high efficiency Mono PERC solar modules and is based on seasonal tilt technology with string invertors. However, solar power plants are practically is not feasible for urban areas where large amount of space is required. India's population growth and rapid urbanization land availability for solar plants will be a complicated issue. Hence, solar rooftops should be given higher priority. Some of the major PV installations include Braboune stadium, Mumbai, which is the world's largest solar rooftop with capacity of 820.8 kWp, as shown in Figure 7a. Another major installation is carport at Cochin International Airport, which is India's largest carport solarized by Tata Power Solar. The plant is 2.67 MW, and is spread across an area of 20289.9 m^2, which offsets 1868 tons of CO_2 as shown in Figure 7b.

(a)   (b)

Figure 7. (**a**) World's largest solar rooftop with a capacity of 820.8 kWp installed on Braboune stadium, at Cricket Club of India, in Mumbai (18.93° N, 72.82° E) (**b**) India's largest solar carport 2.67 Mw at Cochin International Airport (Cial) (10.15° N, 76.39° E).

Although, in India, 83.3 crore reside in a rural area out of 121 crores, urbanization is occuring at a rapid pace. Every minute, 30 Indians move from a rural area to a city, seeking better-paying jobs. Population and economic growth have fostered urbanization in the country and the number of urban towns and cities has drastically increased [146,147]. By the end of 2030, 590 million Indians will be in city for which new buildings are required. It is expected that five-fold built space will be in 2030 than 2005 level in India, of which 60% will be air-conditioned space. Maintaining similar conditions, Indian's building can consume energy and emits GHG with a 700% increment by 2050, compared to 2005 levels [135]. Presently building consumes 30% of electricity in India [148]. The reduction of building energy enhances the demand for sustainable building, which will perform as low energy or less energy-hungry building by trim down the HVAC load demand. To attain such a building, envelopes need to be energy efficient and responsive to an outdoor conundrum. To assess the performance of the buildings, The Energy and Resource Institute of India (TERI) and MNRE has created Green Rating for Integrated Habitat Assessment (GRIHA), and the Leadership in Energy and Environmental Design (LEED) rating tools to help curtail the substantial resources consumed by the building industry, and to reduce the overall environmental impact within tolerable limits. GRIHA evaluates the ecological performance of the building comprehensively by controlling energy consumption, reducing carbon dioxide gas emissions and reinforce the use of inexhaustible and processable sources to the best possible extent [149,150]. MNRE also encourages now passive building which will use solar energy by the suitable orientation of building for daylighting, heating and cooling load demand [151].

BIPV and BAPV both can contribute a considerable amount of energy and improve the building's indoor environment in India [152,153]. However, the dearth of BIPV experts and BIPV marketing professionals, limited in-house consumption data, dearth of ability in planning, commissioning, operation, and maintenance of solar PV/BIPV projects, inadequate training and capacity building, not enough available information about BIPV for policy-making and mobilizing civil society are the barrier for Indian BIPV/BAPV market [81,82]. Another major barrier for widespread PV in India is the lack of resources of raw materials for PV manufacturing. BIPV technology, which is mainly thin-film-based, did not have much uptake in India due to the same reason. For crystalline silicon, India depends on import of the silicon wafer. Currently, in India, the thin-film PV industries are run by US-based First Solar (22% share), Canadian Solar (6% share) and 6% share of Trina Solar Chinese manufacture (6%). India also cannot support CdTe production as India's copper refining industry size is not big enough and upgradation is required to enhance tellurium recovery rates from the copper refining process [154]. The Indian developer, Vikram Solar, has a 3.5% share followed by Moser Baer, Tata Power Solar and Lanco [155]. China controls over 97% of rare earth material which makes them capable to control the price of thin-film [155]. Poor performance of thin-film PV system compared to silicon PV system creates a negative impact on thin-film BIPV system. Thin film degradation occurs in higher rate than crystalline PV over 25 years. Also, thin-film PV cells possess micro-cracks after few years of operation, due to the temperature gradient differences between bottom and top, which cause additional cost of replacement. In India, utilization of BIPV and BAPV is still not fully well established and primarily most of the major integration types are BAPV. Solar rooftop PV application which is BAPV technologies are predominant in India. According to census 2011, in India, there are 331 million households, with urban settlement area of 77,370 km^2, which can be a huge potential of 124 GW for rooftop BAPV to satisfy 40 GW rooftop PV power generation target by 2022. Rooftop PV installations grew at robust pace adding 1,836 MW in the financial year 2018–2019, with a total becoming 10 GW. Figure 8a shows the spaced type semitransparent crystalline solar BIPV module integrated on the rooftop having installed capacity of 1.68 kWp. Each module had dimensions of 1963 mm × 0.987 mm × 40 mm covered with 36 c-Si panels with a transparent area of 49% and rated power of 150 W. Figure 8b shows the India's first zero energy building, which was constructed in 2014. PV panels occupy 4600 m^2 area and annual energy generation cost is 14 lakh ($19k) Unit kWh, while the cost of installation was Rs 18 crore ($2533k). Coal India Limited's Corporate Headquarters at Rajarhat in Kolkata (22.57° N, 88.37° E) installed 632 PV panels with total capacity of 140 kW. The solar energy powers the uninterrupted power supply for desktops, emergency lighting systems and the landscape lighting of CIL's corporate office. Tata BP Solar has implemented 30 kWp ($250,000) BAPV system in Samudra Institute of Maritime Studies in Pune (18.52° N, 73.85° E). Moser Baer has installed a 1.8 kWp BIPV exterior façade of Jubilee shopping complex in Hyderabad (17.38° N, 78.48° E) to meet power requirements in shopping complexes. The government buildings in India are also encouraged to use solar energy in an aesthetic approach by using BIPV/BAPV technology. The government is also focused on increasing the roof top systems and streamlining policy implementation processes. In 2015, Novus green installed a 1MW BAPV system (4000 PV system each had 250 Wp) at the rooftop of IIT Delhi. Energy Efficiency Services (EESL) had planned to invest INR 800 crore for rooftop solar in Maharashtra across 5000 state-owned buildings to install 200 MW grid-connected systems. EESL estimates that about 100 million units would be saved per year by replacing energy-inefficient ceiling fans (6 lakh), LED bulbs (11 lakh) and air-conditioners (7000) along with streetlights (14,000) and retrofitting 3000 buildings.

(a)   (b)

Figure 8. (**a**) Spaced type crystalline silicon solar-based BIPV roof for daylighting application (Source: HHV solar, Bangalore, India), (**b**) BAPV system in Indira Paryavaran Bhawan India (Image source: BT).

Another barrier in India for poor BIPV/BAPV - standalone system is the unorganized Indian electricity market. The PV power generation sector includes three different customers. The first customer state distribution companies (DISCOMs) who have renewable purchase obligations (RPO) to buy PV power and meet 10.5% PV electricity generated, second is the rooftop PV consumers (RPVCs) and third is large buyers of power who are also known as open access consumers (OACs) [156]. Under open access (OA), consumers are capable of buying electricity from producers who generate electricity independently. Indian rail started exploring PV power from OA. Developers feel that RPO is not same for all state as State Electricity Regulatory Commissions (SERCs) has different benchmark for each state. PV power electricity price varies from INR 7.5 to INR18.5/kWh. In 2010, Central Electricity Regularity Commission (CERC) has implemented the PV/BIPV feed-in tariff of INR 17.9/kWh [157]. In India, the coal power electricity price is about INR 5.5/kW h, while the PV power price is about INR6.5/kW h. Hence, project developers are compelled to offer discounts [158]. Thus, DISCOMs are failing to comply with RPO requirements, due to their poor financial health and lower solar tariff support. Hence, they prefer to wait for buying low price PV electricity. DISCOMs also argued that BIPV/BAPV based standalone systems increase their financial burden as RPV customers prefer to buy grid electricity over BIPV/BAPV power due to intermittency. Intermittent PV power generation is also an issue for the promotion of OA [159]. To rectify this issue, energy banking can be created where DISCOM will facilitate OA transactions through electricity banking, between an independent power producer and OAC. DISCOM can generate less power whenever an independent power producer generates a higher amount of power than the OAC's demand, in order to use surplus PV power and generate when PV generation is reduced and shortfalls arise [159,160]. The presence of multiple electricity regulatory boards like MNRE (Ministry of New and Renewable Energy), CERC (Central Electricity Regulatory Commission), SERC (Ministry of Power and State Electricity Regulatory Commission) also make the legal process a cumbersome task [161].

By the end of Sep 2019, India's cumulative installed solar capacity stood at 33.8 GW, of which 88% are PV plants and 12% rooftop installations. Ambiguity in incentive implementation, non-availability of storage systems incentives, lack of consumer awareness and research studies are the reason behind this sluggish movement for rooftop PV application in India [162]. Dust accumulation, which reduces PV power generation, should also be taken care of as air quality in India can pose a negative impact [163–165].

5. Perspective and Discussion

It is evident that in India, solar PV power generation got heads up after the year of 2015. Rooftop installation got priority however it included rooftop of any large construction whether it is a building (residential/commercial) or other area, such as a stadium or car park. Hence, it is difficult to differentiate or estimate the installed percentage for only rooftop building. Until now, power generation from PV has only just become a higher priority than aesthetic application. In India, BAPV system is prevalent and they are wrongly termed as BIPV systems. Actual, BIPV technology is not particularly popular in India, which could be the reason for lack of government plan and policy, and awareness to the public. BIPV window tiles or foil technology is not popular in India. Thermal regulation of BAPVT in the name of BIPVT is available in India [129]. The primary goal of this work to enhance PV power generation, rather than reducing the building cooling load demand. Improvements in the building environment, using BIPV, should be in government policy. Concentrating PV in India is mainly higher in concentrators [166]. To the best of our knowledge, no work has been reported on a low concentrator using LSC- or CPC-based BIPV/BAPV in India. To meet the GRIHA rating, large commercial buildings are now growing, however, they do not use BIPV to improve the built environment. Most green buildings, which meet the GRIHA rating, use rooftop BAPV system, while external shading devices control the admitted daylight and heat for those large glazed façades [167]. Table 3 listed the availability of BIPV and BAPV products in India.

Table 3. Availability of BIPV/BAPV products in India.

Product	Availability in Research Paper	Reference
BIPV window	×	
BIPV foil	×	
BIPV tile	×	
BAPVT (water/air)	✓	[168,169]
BIPV/BAPV-(PCM)	✓	[170,171]
BIPV/BAPV-LSC	×	
BIPV/BAPV-CPC	×	

Future Pathway of BIPV/BAPV in India

Weak BIPV implementation and national planning, lack of energy policy and details of BIPV products, fewer BIPV experts and market professionals are the key responsible factors for slow or no progress of BIPV, for less-energy hungry building in India. The Indian Central government should motivate and support deployment of BIPV research and development by removing non-economic issues to BIPV uses, creating building codes for BIPV integration in building assembly. New training programs related to BIPV can educate the builders, developers, and engineers. Support from central or state-level government organizations, such as NISE, SECI, NIWE, MNRE, IIT, NITs and the state nodal agencies should work together.

Windows are one of the weakest components in a building. It allows external heat to come inside (enhance the air condition load), internal heat to outside (enhance heating load) and offer visual connection to building interior to exterior [172–185]. Building window systems are affected by an overall heat transfer coefficient (U-value) and solar energy transmittance (g-value) [186–189]. For warmer place, high U-value and low g-value are required, while for colder area, low U-value and high g-value are suitable. Generally, single pane glass possesses higher U-value (U-value 3–5 W/m^{2}K) and higher g-value followed by double (U-value-2-3 W/m^{2}K g-value lower than single glass) and triple (U-value < 2W/m^{2}K; g-value lower than single and double) glass window [190–193]. This clear and highly transparent window is not able to limit the heat entering from exterior ambient of buildings. Thus, building interior temperature often crosses over occupants' comfort limit (thermal comfort temperature 18–20 °C). Hence, in order to maintain thermal comfort level, an excessive amount of grid power is consumed to run air-condition (AC). Integration of PV system into the single or double glass can create a single glass BIPV window or double glass BIPV window, which will, not only generate benevolent electricity, but also contro heat and restrict its flow into the exterior, where required, as shown in Figure 9. Semitransparent c-Si PV based BIPV windows can reduce 5.3% heating and cooling load compared to standard BIPV [194] and has ability to limit up to 65% in total heat gains compared to traditional clear glass [195]. Previous investigation of BIPV window in cooling load dominated climate such as Singapore, Hongkong showed a positive impact on load reduction [196–198].

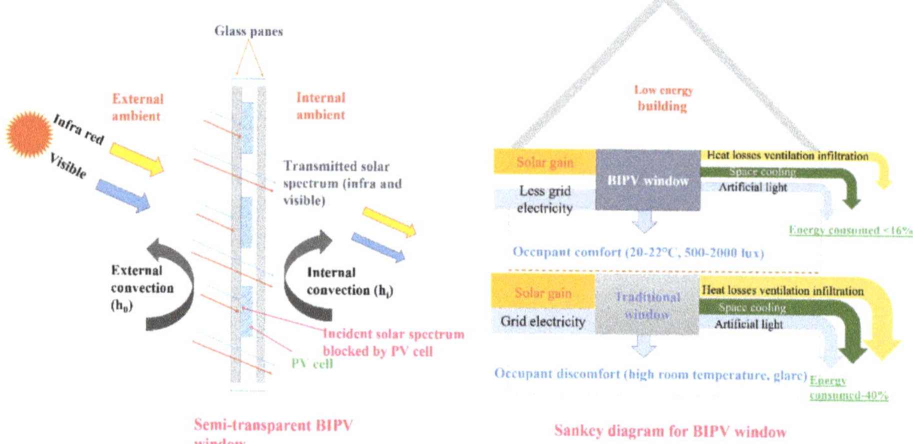

Figure 9. Schematic of a semitransparent BIPV window (**left**) and Sankey diagram while BIPV window is integrated into a building (**right**).

In India, buildings' AC load is excessively high, due to thehigh g-value of window. In summer, buildings' windows are closed, in order to abate hot air and sunlight [199,200]. This also creates a dearth of light in an indoor setting, which encourages occupants to employ artificial light. Small-to-medium office buildings use air conditioners during the day and peak summer, and are not in use at night or off-peak season. Hence, advanced single or double glass-based semitransparent BIPV window systems, which possess lower g-value compared to clear glass, are favorable in India as they can be particularly be influential in limiting excess usage of AC and lighting load for a less-energy hungry building.The inclusion of this semi-transparent BIPV window, not only lowers the g-value, but allows sufficient daylight and generates benign electricity concomitantly. India has primarily cooling load (AC load) demand climate, but can trim down this excessive grid power consumption by employing BIPV window to obtain less energy-hungry building. Also, replacing the traditional window system by BIPV window is easier than replacing other building components, such as roof or wall [201,202].

6. Conclusions

The following conclusion can be drawn from this review article:

- Building integrated photovoltaic (BIPV) replaces traditional building envelop, such as window, wall, roof and most often they are thin film, or third-generation based transparent or semi-transparent in nature.
- Building attached/applied photovoltaic (BAPV) indicates when PV systems are attached to a building without replacing its traditional envelops.
- India's solar mission which geared up from 2015, accelerated the rooftop PV integration. For building, the rooftop application's majority are BAPV types and capacity higher than kW level. India's total installed solar capacity reached only 33.8 GW by the end of September 2019, which is way behind to achieve India's 40 GW rooftop PV power generation by 2022.
- New technologies, such as PV tiles, foil and windows, as part of BIPV or BPAV, are not popular in India.

- Indian electricity market needs a complete reform to allow smooth penetration of BIPV/BAPV in building. Also, due to rapid urbanization and development of zero and sustainable building industry, BIPV will keep pace in India soon.
- Single and double glass BIPV window systems have been identified as one of the most potential candidates for India.

Author Contributions: Conceptualization, A.G.; methodology, A.G.; investigation, P.R., M.V.N.S.G., A.K., S.N., A.G.; resources, P.R., S.N., A.G.; writing—original draft preparation, P.R., A.G.; writing—review and editing, S.N., A.G.; supervision, A.G.; project administration, A.G.; funding acquisition, A.G. All authors have read and agreed to the published version of the manuscript.

Funding: This work didn't receive any specific grants.

Acknowledgments: The authors would like to thank the anonymous reviewers for their valuable comments which were useful to enhance the quality of paper.

Conflicts of Interest: The authors declare no conflict of interest.

References

1. Kumar Sahu, B. A study on global solar PV energy developments and policies with special focus on the top ten solar PV power producing countries. *Renew. Sustain. Energy Rev.* **2015**, *43*, 621–634. [CrossRef]
2. Shukla, A.K.; Sudhakar, K.; Baredar, P. A comprehensive review on design of building integrated photovoltaic system. *Energy Build.* **2016**, *128*, 99–110. [CrossRef]
3. Gupta, D.; Ghersi, F.; Vishwanathan, S.S.; Garg, A. Achieving sustainable development in India along low carbon pathways: Macroeconomic assessment. *World Dev.* **2019**, *123*, 104623. [CrossRef]
4. Muniyoor, K. Is there a trade-off between energy consumption and employment: Evidence from India. *J. Clean. Prod.* **2020**, *255*, 120262. [CrossRef]
5. Tang, X.; McLellan, B.C.; Zhang, B.; Snowden, S.; Höök, M. Trade-Off analysis between embodied energy exports and employment creation in China. *J. Clean. Prod.* **2016**, *134*, 310–319. [CrossRef]
6. Bano, F.; Kamal, M.A. Examining the role of building envelope for energy efficiency in office buildings in India. *Archit. Res.* **2016**, *6*, 107–115. [CrossRef]
7. Bhatt, M.S.; Rajkumar, N.; Jothibasu, S.; Sudirkumar, R.; Pandian, G.; Nair, K.R.C. Commercial and residential building energy labeling. *J. Sci. Ind. Res. (India)* **2005**, *64*, 30–34.
8. Manu, S.; Brager, G.; Rawal, R.; Geronazzo, A.; Kumar, D. Performance evaluation of climate responsive buildings in India—Case studies from cooling dominated climate zones. *Build. Environ.* **2019**, *148*, 136–156. [CrossRef]
9. Bandyopadhyay, S.; Pathak, C.R.; Dentinho, T.P. *Urbanization and Regional Sustainability in South Asia*; Springer Open Ltd.: Cham, Switzerland, 2020; ISBN 978-3-030-23795-0.
10. Zhang, S.; He, Y. Analysis on the development and policy of solar PV power in China. *Renew. Sustain. Energy Rev.* **2013**, *21*, 393–401. [CrossRef]
11. Thapar, S.; Sharma, S. Factors impacting wind and solar power sectors in India: A survey-based analysis. *Sustain. Prod. Consum.* **2020**, *21*, 204–215. [CrossRef]
12. Righini, G.C.; Enrichi, F. *Solar Cells' Evolution and Perspectives: A Short Review*; Elsevier Ltd.: Amsterdam, The Netherlands, 2020; ISBN 9780081027622.
13. Røyset, A.; Kolås, T.; Jelle, B.P. Coloured building integrated photovoltaics: Influence on energy efficiency. *Energy Build.* **2020**, *208*. [CrossRef]
14. Peharz, G.; Ulm, A. Quantifying the influence of colors on the performance of c-Si photovoltaic devices. *Renew. Energy* **2018**, *129*, 299–308. [CrossRef]
15. Saga, T. Advances in crystalline silicon solar cell technology for industrial mass production. *NPG Asia Mater.* **2010**, *2*, 96–102. [CrossRef]
16. Liu, Z.; Sofia, S.E.; Laine, H.S.; Woodhouse, M.; Wieghold, S.; Peters, I.M.; Buonassisi, T. Revisiting thin silicon for photovoltaics: A technoeconomic perspective. *arXiv* **2019**, arXiv:1906.06770. [CrossRef]
17. Battaglia, C.; Cuevas, A.; De Wolf, S. High-Efficiency crystalline silicon solar cells: Status and perspectives. *Energy Environ. Sci.* **2016**, *9*, 1552–1576. [CrossRef]

18. Chen, Y.; Altermatt, P.P.; Chen, D.; Zhang, X.; Xu, G.; Yang, Y.; Wang, Y.; Feng, Z.; Shen, H.; Verlinden, P.J. From laboratory to production: Learning models of efficiency and manufacturing cost of industrial crystalline silicon and thin-film photovoltaic technologies. *IEEE J. Photovolt.* **2018**, *8*, 1531–1538. [CrossRef]

19. Ardente, F.; Latunussa, C.F.L.; Blengini, G.A. Resource efficient recovery of critical and precious metals from waste silicon PV panel recycling. *Waste Manag.* **2019**, *91*, 156–167. [CrossRef]

20. Hughes, L.; Bristow, N.; Korochkina, T.; Sanchez, P.; Gomez, D.; Kettle, J.; Gethin, D. Assessing the potential of steel as a substrate for building integrated photovoltaic applications. *Appl. Energy* **2018**, *229*, 209–223. [CrossRef]

21. Chen, M.; Zhang, W.; Xie, L.; Ni, Z.; Wei, Q.; Wang, W.; Tian, H. Experimental and numerical evaluation of the crystalline silicon PV window under the climatic conditions in southwest China. *Energy* **2019**, *183*, 584–598. [CrossRef]

22. Akhsassi, M.; El Fathi, A.; Erraissi, N.; Aarich, N.; Bennouna, A.; Raoufi, M.; Outzourhit, A. Experimental investigation and modeling of the thermal behavior of a solar PV module. *Sol. Energy Mater. Sol. Cells* **2018**, *180*, 271–279. [CrossRef]

23. Gottschalg, R.; Betts, T.R.; Eeles, A.; Williams, S.R.; Zhu, J. Influences on the energy delivery of thin film photovoltaic modules. *Sol. Energy Mater. Sol. Cells* **2013**, *119*, 169–180. [CrossRef]

24. Taraba, M.; Adamec, J.; Danko, M.; Drgona, P.; Urica, T. Properties measurement of the thin film solar panels under adverse weather conditions. *Transp. Res. Procedia* **2019**, *40*, 535–540. [CrossRef]

25. Shah, A.V.; Schade, H.; Vanecek, M.; Meier, J.; Vallat-Sauvain, E.; Wyrsch, N.; Kroll, U.; Droz, C.; Bailat, J. Thin-Film silicon solar cell technology. *Photovolt. Res. Appl.* **2004**, *12*, 113–142. [CrossRef]

26. Sebastian, P.J.; Sivaramakrishnan, V. Electrical conduction and transmission electron microscopy studies of $CdSe_{0.8}Te_{0.2}$ thin films. *Thin Solid Films* **1991**, *202*, 1–9. [CrossRef]

27. Lee, T.D.; Ebong, A.U. A review of thin film solar cell technologies and challenges. *Renew. Sustain. Energy Rev.* **2017**, *70*, 1286–1297. [CrossRef]

28. Matsui, T.; Sai, H.; Bidiville, A.; Hsu, H.J.; Matsubara, K. Progress and limitations of thin-film silicon solar cells. *Sol. Energy* **2018**, *170*, 486–498. [CrossRef]

29. Stuckelberger, M.; Biron, R.; Wyrsch, N.; Haug, F.-J.; Ballif, C. Review: Progress in solar cells from hydrogenated amorphous silicon. *Renew. Sustain. Energy Rev.* **2017**, *76*, 1–27. [CrossRef]

30. Myong, S.Y.; Jeon, S.W. Efficient outdoor performance of esthetic bifacial a-Si: H semi-transparent PV modules. *Appl. Energy* **2016**, *164*, 312–320. [CrossRef]

31. Kichou, S.; Silvestre, S.; Nofuentes, G.; Torres-Ramírez, M.; Chouder, A.; Guasch, D. Characterization of degradation and evaluation of model parameters of amorphous silicon photovoltaic modules under outdoor long term exposure. *Energy* **2016**, *96*, 231–241. [CrossRef]

32. Sabri, L.; Benzirar, M. Effect of ambient conditions on thermal properties of photovoltaic cells: Crystalline and amorphous silicon. *Int. J. Innov. Res. Sci. Eng. Technol.* **2014**, *3*, 17815–17821. [CrossRef]

33. Staebler, D.L.; Wronski, C.R. Reversible conductivity changes in discharged-produced amorphous Si. *Appl. Phys. Lett.* **1977**, *31*, 274–292. [CrossRef]

34. Feurer, T.; Carron, R.; Torres Sevilla, G.; Fu, F.; Pisoni, S.; Romanyuk, Y.E.; Buecheler, S.; Tiwari, A.N. Efficiency improvement of near-stoichiometric $CuInSe_2$ solar cells for application in tandem devices. *Adv. Energy Mater.* **2019**, *9*, 2–7. [CrossRef]

35. Kazmerski, L.L.; White, F.R.; Morgan, G.K. Thin-Film $CuInSe_2$/CdS heterojunction solar cells. *Appl. Phys. Lett.* **1976**, *29*, 268–270. [CrossRef]

36. Stoichkov, V.; Sweet, T.K.N.; Jenkins, N.; Kettle, J. Studying the outdoor performance of organic building-integrated photovoltaics laminated to the cladding of a building prototype. *Sol. Energy Mater. Sol. Cells* **2019**, *191*, 356–364. [CrossRef]

37. Sato, R.; Chiba, Y.; Chikamatsu, M.; Yoshida, Y.; Taima, T.; Kasu, M.; Masuda, A. Investigation of the power generation of organic photovoltaic modules connected to the power grid for more than three years. *Jpn. J. Appl. Phys.* **2019**, *58*, 52001. [CrossRef]

38. Chemisana, D.; Moreno, A.; Polo, M.; Aranda, C.; Riverola, A.; Ortega, E.; Lamnatou, C.; Domènech, A.; Blanco, G.; Cot, A. Performance and stability of semitransparent OPVs for building integration: A benchmarking analysis. *Renew. Energy* **2019**, *137*, 177–188. [CrossRef]

39. O'Regan, B.; Gratzel, M. A low-cost, high efficiency solar cell based on dye-sensitized colloidal TiO_2 films. *Nature* **1991**, *353*, 737–740. [CrossRef]

40. Sharma, S.; Siwach, B.; Ghoshal, S.K.; Mohan, D. Dye sensitized solar cells: From genesis to recent drifts. *Renew. Sustain. Energy Rev.* **2017**, *70*, 529–537. [CrossRef]
41. Kumara, N.T.R.N.; Lim, A.; Lim, C.M.; Petra, M.I.; Ekanayake, P. Recent progress and utilization of natural pigments in dye sensitized solar cells: A review. *Renew. Sustain. Energy Rev.* **2017**, *78*, 301–317. [CrossRef]
42. Ghosh, A.; Selvaraj, P.; Sundaram, S.; Mallick, T.K. The colour rendering index and correlated colour temperature of dye-sensitized solar cell for adaptive glazing application. *Sol. Energy* **2018**, *163*, 537–544. [CrossRef]
43. Cornaro, C.; Renzi, L.; Pierro, M.; Di Carlo, A.; Guglielmotti, A. Thermal and electrical characterization of a semi-transparent dye-sensitized photovoltaic module under real operating conditions. *Energies* **2018**, *11*, 155. [CrossRef]
44. Ghosh, A.; Bhandari, S.; Sundaram, S.; Mallick, T.K. Carbon counter electrode mesoscopic ambient processed & characterised perovskite for adaptive BIPV fenestration. *Renew. Energy* **2020**, *145*, 2151–2158. [CrossRef]
45. Bhandari, S.; Roy, A.; Ghosh, A.; Mallick, T.K.; Sundaram, S. Performance of WO$_3$—Incorporated carbon electrodes for ambient mesoscopic perovskite solar cells. *ACS Omega* **2019**, *5*, 422–429. [CrossRef] [PubMed]
46. Cannavale, A.; Ierardi, L.; Hörantner, M.; Eperon, G.E.; Snaith, H.J.; Ayr, U.; Martellotta, F. Improving energy and visual performance in offices using building integrated perovskite-based solar cells: A case study in Southern Italy. *Appl. Energy* **2017**, *205*, 834–846. [CrossRef]
47. Cannavale, A.; Hörantner, M.; Eperon, G.E.; Snaith, H.J.; Fiorito, F.; Ayr, U.; Martellotta, F. Building integration of semitransparent perovskite-based solar cells: Energy performance and visual comfort assessment. *Appl. Energy* **2017**, *194*, 94–107. [CrossRef]
48. Chen, Y.; Chen, D.; Liu, C.; Wang, Z.; Zou, Y.; He, Y.; Wang, Y.; Yuan, L.; Gong, J.; Lin, W.; et al. Mass production of industrial tunnel oxide passivated contacts (i-TOPCon) silicon solar cells with average efficiency over 23% and modules over 345 W. *Prog. Photovolt. Res. Appl.* **2019**, 1–8. [CrossRef]
49. Glunz, S.W.; Preu, R.; Biro, D. *Crystalline Silicon Solar Cells: State-of-the-Art and Future Developments*; Elsevier Ltd.: Amsterdam, The Netherlands, 2012; Volume 1, ISBN 9780080878737.
50. Zhang, T.; Wang, M.; Yang, H. A review of the energy performance and life-cycle assessment of building-integrated photovoltaic (BIPV) systems. *Energies* **2018**, *11*, 3157. [CrossRef]
51. Haegel, N.M.; Atwater, H.; Barnes, T.; Breyer, C.; Burrell, A.; Chiang, Y.-M.; De Wolf, S.; Dimmler, B.; Feldman, D.; Glunz, S.; et al. Terawatt-Scale photovoltaics: Transform global energy. *Science* **2019**, *364*, 836–838. [CrossRef]
52. Peng, J.; Lu, L.; Yang, H. Review on life cycle assessment of energy payback and greenhouse gas emission of solar photovoltaic systems. *Renew. Sustain. Energy Rev.* **2013**, *19*, 255–274. [CrossRef]
53. Tsai, C.Y.; Tsai, C.Y. See-through, light-through, and color modules for large-area tandem amorphous/microcrystalline silicon thin-film solar modules: Technology development and practical considerations for building-integrated photovoltaic applications. *Renew. Energy* **2020**, *145*, 2637–2646. [CrossRef]
54. Minemoto, T.; Fukushige, S.; Takakura, H. Difference in the outdoor performance of bulk and thin-film silicon-based photovoltaic modules. *Sol. Energy Mater. Sol. Cells* **2009**, *93*, 1062–1065. [CrossRef]
55. Han, J.; Lu, L.; Yang, H. Numerical evaluation of the mixed convective heat transfer in a double-pane window integrated with see-through a-Si PV cells with low-e coatings. *Appl. Energy* **2010**, *87*, 3431–3437. [CrossRef]
56. Zidane, T.E.K.; Adzman, M.R.B.; Tajuddin, M.F.N.; Mat Zali, S.; Durusu, A. Optimal configuration of photovoltaic power plant using grey wolf optimizer: A comparative analysis considering CdTe and c-Si PV modules. *Sol. Energy* **2019**, *188*, 247–257. [CrossRef]
57. Kumar, N.M.; Sudhakar, K.; Samykano, M. Performance evaluation of CdTe BIPV roof and façades in tropical weather conditions. *Energy Sources Part A Recover. Util. Environ. Eff.* **2019**, 1–15. [CrossRef]
58. Bosio, A.; Rosa, G.; Romeo, N. Past, present and future of the thin film CdTe/CdS solar cells. *Sol. Energy* **2018**, 31–43. [CrossRef]
59. Rawat, R.; Kaushik, S.C.; Sastry, O.S.; Bora, B.; Singh, Y.K. Long-Term performance analysis of CdTe PV module in real operating conditions. *Mater. Today Proc.* **2018**, *5*, 23210–23217. [CrossRef]
60. Barman, S.; Chowdhury, A.; Mathur, S.; Mathur, J. Assessment of the efficiency of window integrated CdTe based semi-transparent photovoltaic module. *Sustain. Cities Soc.* **2018**, *37*, 250–262. [CrossRef]
61. Ozden, T.; Akinoglu, B.G.; Turan, R. Long term outdoor performances of three different on-grid PV arrays in central Anatolia—An extended analysis. *Renew. Energy* **2017**, *101*, 182–195. [CrossRef]

62. Powalla, M.; Dimmler, B. Development of large-area CIGS modules. *Sol. Energy Mater. Sol. Cells* **2003**, *75*, 27–34. [CrossRef]

63. Dhere, N.G. Scale-Up issues of CIGS thin film PV modules. *Sol. Energy Mater. Sol. Cells* **2011**, *95*, 277–280. [CrossRef]

64. Yalçin, L.; Öztürk, R. Performance comparison of c-Si, mc-Si and a-Si thin film PV by PVsyst simulation. *J. Optoelectron. Adv. Mater.* **2013**, *15*, 326–334. [CrossRef]

65. De Wild-Scholten, M.J. Energy payback time and carbon footprint of commercial photovoltaic systems. *Sol. Energy Mater. Sol. Cells* **2013**, *119*, 296–305. [CrossRef]

66. Theelen, M.; Foster, C.; Steijvers, H.; Barreau, N.; Vroon, Z.; Zeman, M. The impact of atmospheric species on the degradation of CIGS solar cells. *Sol. Energy Mater. Sol. Cells* **2015**, *141*, 49–56. [CrossRef]

67. Theelen, M.; Daume, F. Stability of Cu(In,Ga)Se$_2$ solar cells: A literature review. *Sol. Energy* **2016**, *133*, 586–627. [CrossRef]

68. Rudra, S.; Sarker, S.; Kim, D.M. Review on simulation of current-voltage characteristics of dye-sensitized solar cells. *J. Ind. Eng. Chem.* **2019**, *80*, 516–526. [CrossRef]

69. Gong, J.; Sumathy, K.; Qiao, Q.; Zhou, Z. Review on dye-sensitized solar cells (DSSCs): Advanced techniques and research trends. *Renew. Sustain. Energy Rev.* **2017**, *68*, 234–246. [CrossRef]

70. Mehmood, U.; Al-Ahmed, A.; Al-Sulaiman, F.A.; Malik, M.I.; Shehzad, F.; Khan, A.U.H. Effect of temperature on the photovoltaic performance and stability of solid-state dye-sensitized solar cells: A review. *Renew. Sustain. Energy Rev.* **2017**, *79*, 946–959. [CrossRef]

71. Howard, I.A.; Abzieher, T.; Hossain, I.M.; Eggers, H.; Schackmar, F.; Ternes, S.; Richards, B.S.; Lemmer, U.; Paetzold, U.W. Coated and printed perovskites for photovoltaic applications. *Adv. Mater.* **2019**, *31*, 1806702. [CrossRef]

72. Iqbal, T.; Shabbir, U.; Sultan, M.; Tahir, M.B.; Farooq, M.; Mansha, M.S.; Ijaz, M.; Maraj, M. All ambient environment-based perovskite film fabrication for photovoltaic applications. *Int. J. Energy Res.* **2019**, *43*, 806–813. [CrossRef]

73. Agresti, A.; Pescetelli, S.; Palma, A.L.; Martín-García, B.; Najafi, L.; Bellani, S.; Moreels, I.; Prato, M.; Bonaccorso, F.; Di Carlo, A. Two-Dimensional material interface engineering for efficient perovskite large-area modules. *ACS Energy Lett.* **2019**, 1862–1871. [CrossRef]

74. Ballif, C.; Perret-Aebi, L.E.; Lufkin, S.; Rey, E. Integrated thinking for photovoltaics in buildings. *Nat. Energy* **2018**, *3*, 438–442. [CrossRef]

75. Peng, C.; Huang, Y.; Wu, Z. Building-Integrated photovoltaics (BIPV) in architectural design in China. *Energy Build.* **2011**, *43*, 3592–3598. [CrossRef]

76. Jelle, B.P.; Breivik, C.; Røkenes, H.D. Building integrated photovoltaic products: A state-of-the-art review and future research opportunities. *Sol. Energy Mater. Sol. Cells* **2012**, *100*, 69–96. [CrossRef]

77. Jelle, B.P. Building integrated photovoltaics: A concise description of the current state of the art and possible research pathways. *Energies* **2016**, *9*, 21. [CrossRef]

78. Shukla, A.K.; Sudhakar, K.; Baredar, P. Exergetic assessment of BIPV module using parametric and photonic energy methods: A review. *Energy Build.* **2016**, *119*, 62–73. [CrossRef]

79. Shukla, A.K.; Sudhakar, K.; Baredar, P. Recent advancement in BIPV product technologies: A review. *Energy Build.* **2017**, *140*, 188–195. [CrossRef]

80. Shukla, A.K.; Sudhakar, K.; Baredar, P.; Mamat, R. BIPV in Southeast Asian countries—Opportunities and challenges. *Renew. Energy Focus* **2017**, *21*, 25–32. [CrossRef]

81. Shukla, A.K.; Sudhakar, K.; Baredar, P.; Mamat, R. BIPV based sustainable building in South Asian countries. *Sol. Energy* **2018**, *170*, 1162–1170. [CrossRef]

82. Shukla, A.K.; Sudhakar, K.; Baredar, P.; Mamat, R. Solar PV and BIPV system: Barrier, challenges and policy recommendation in India. *Renew. Sustain. Energy Rev.* **2018**, *82*, 3314–3322. [CrossRef]

83. Ghosh, A.; Sundaram, S.; Mallick, T.K. Investigation of thermal and electrical performances of a combined semi-transparent PV-vacuum glazing. *Appl. Energy* **2018**, *228*, 1591–1600. [CrossRef]

84. Ghosh, A.; Sundaram, S.; Mallick, T.K. Colour properties and glazing factors evaluation of multicrystalline based semi-transparent Photovoltaic-vacuum glazing for BIPV application. *Renew. Energy* **2019**, *131*, 730–736. [CrossRef]

85. Wang, M.; Peng, J.; Li, N.; Yang, H.; Wang, C.; Li, X.; Lu, T. Comparison of energy performance between PV double skin facades and PV insulating glass units. *Appl. Energy* **2017**, *194*, 148–160. [CrossRef]

86. Peng, J.; Curcija, D.C.; Lu, L.; Selkowitz, S.E.; Yang, H.; Zhang, W. Numerical investigation of the energy saving potential of a semi-transparent photovoltaic double-skin facade in a cool-summer Mediterranean climate. *Appl. Energy* **2016**, *165*, 345–356. [CrossRef]

87. Zhang, W.; Lu, L.; Peng, J.; Song, A. Comparison of the overall energy performance of semi-transparent photovoltaic windows and common energy-efficient windows in Hong Kong. *Energy Build.* **2016**, *128*, 511–518. [CrossRef]

88. Peng, J.; Lu, L.; Yang, H.; Ma, T. Comparative study of the thermal and power performances of a semi-transparent photovoltaic façade under different ventilation modes. *Appl. Energy* **2015**, *138*, 572–583. [CrossRef]

89. Alrashidi, H.; Ghosh, A.; Issa, W.; Sellami, N.; Mallick, T.K.; Sundaram, S. Evaluation of solar factor using spectral analysis for CdTe photovoltaic glazing. *Mater. Lett.* **2019**, *237*, 332–335. [CrossRef]

90. Alrashidi, H.; Ghosh, A.; Issa, W.; Sellami, N.; Mallick, T.K.; Sundaram, S. Thermal performance of semitransparent CdTe BIPV window at temperate climate. *Sol. Energy* **2020**, *195*, 536–543. [CrossRef]

91. Selvaraj, P.; Ghosh, A.; Mallick, T.K.; Sundaram, S. Investigation of semi-transparent dye-sensitized solar cells for fenestration integration. *Renew. Energy* **2019**, *141*, 516–525. [CrossRef]

92. Roy, A.; Ghosh, A.; Bhandari, S.; Selvaraj, P.; Sundaram, S.; Mallick, T.K. Color comfort evaluation of dye-sensitized solar cell (DSSC) based building-integrated photovoltaic (BIPV) glazing after 2 years of ambient exposure. *J. Phys. Chem. C* **2019**, *123*, 23834–23837. [CrossRef]

93. Sun, Y.; Shanks, K.; Baig, H.; Zhang, W.; Hao, X.; Li, Y.; He, B.; Wilson, R.; Liu, H.; Sundaram, S.; et al. Integrated semi-transparent cadmium telluride photovoltaic glazing into windows: Energy and daylight performance for different architecture designs. *Appl. Energy* **2018**, *231*, 972–984. [CrossRef]

94. Li, G.; Xuan, Q.; Akram, M.W.; Golizadeh Akhlaghi, Y.; Liu, H.; Shittu, S. Building integrated solar concentrating systems: A review. *Appl. Energy* **2020**, *260*, 114288. [CrossRef]

95. Rafiee, M.; Chandra, S.; Ahmed, H.; McCormack, S.J. An overview of various configurations of luminescent solar concentrators for photovoltaic applications. *Opt. Mater.* **2019**, *91*, 212–227. [CrossRef]

96. Baig, H.; Sellami, N.; Mallick, T.K. Performance modeling and testing of a building integrated concentrating photovoltaic (BICPV) system. *Sol. Energy Mater. Sol. Cells* **2015**, *134*, 29–44. [CrossRef]

97. Tian, M.; Su, Y.; Zheng, H.; Pei, G.; Li, G.; Riffat, S. A review on the recent research progress in the compound parabolic concentrator (CPC) for solar energy applications. *Renew. Sustain. Energy Rev.* **2018**, *82*, 1272–1296. [CrossRef]

98. Shanks, K.; Senthilarasu, S.; Mallick, T.K. Optics for concentrating photovoltaics: Trends, limits and opportunities for materials and design. *Renew. Sustain. Energy Rev.* **2016**, *60*, 394–407. [CrossRef]

99. Baig, H.; Heasman, K.C.; Mallick, T.K. Non-Uniform illumination in concentrating solar cells. *Renew. Sustain. Energy Rev.* **2012**, *16*, 5890–5909. [CrossRef]

100. Sarmah, N.; Mallick, T.K. Design, fabrication and outdoor performance analysis of a low concentrating photovoltaic system. *Sol. Energy* **2015**, *112*, 361–372. [CrossRef]

101. Sellami, N.; Mallick, T.K. Optical characterisation and optimisation of a static window integrated concentrating photovoltaic system. *Sol. Energy* **2013**, *91*, 273–282. [CrossRef]

102. Mallick, T.K.; Eames, P.C.; Hyde, T.J.; Norton, B. The design and experimental characterisation of an asymmetric compound parabolic photovoltaic concentrator for building façade integration in the UK. *Sol. Energy* **2004**, *77*, 319–327. [CrossRef]

103. Muhammad-Sukki, F.; Abu-Bakar, S.H.; Ramirez-Iniguez, R.; McMeekin, S.G.; Stewart, B.G.; Sarmah, N.; Mallick, T.K.; Munir, A.B.; Mohd Yasin, S.H.; Abdul Rahim, R. Mirror symmetrical dielectric totally internally reflecting concentrator for building integrated photovoltaic systems. *Appl. Energy* **2014**, *113*, 32–40. [CrossRef]

104. Selvaraj, P.; Baig, H.; Mallick, T.K.; Siviter, J.; Montecucco, A.; Li, W.; Paul, M.; Sweet, T.; Gao, M.; Knox, A.R.; et al. Enhancing the efficiency of transparent dye-sensitized solar cells using concentrated light. *Sol. Energy Mater. Sol. Cells* **2018**, *175*, 29–34. [CrossRef]

105. Baig, H.; Kanda, H.; Asiri, A.M.; Nazeeruddin, M.K.; Mallick, T. Increasing efficiency of perovskite solar cells using low concentrating photovoltaic systems. *Sustain. Energy Fuels* **2020**, 1–10. [CrossRef]

106. Chandra, S.; Doran, J.; McCormack, S.J.; Kennedy, M.; Chatten, A.J. Enhanced quantum dot emission for luminescent solar concentrators using plasmonic interaction. *Sol. Energy Mater. Sol. Cells* **2012**, *98*, 385–390. [CrossRef]

107. Chandra, S.; Doran, J.; McCormack, S.J. Two step continuous method to synthesize colloidal spheroid gold nanorods. *J. Colloid Interface Sci.* **2015**, *459*, 218–223. [CrossRef] [PubMed]

108. Chandra, S.; McCormack, S.J.; Kennedy, M.; Doran, J. Quantum dot solar concentrator: Optical transportation and doping concentration optimization. *Sol. Energy* **2015**, *115*, 552–561. [CrossRef]

109. Ter Schiphorst, J.; Cheng, M.L.M.K.H.Y.K.; van der Heijden, M.; Hageman, R.L.; Bugg, E.L.; Wagenaar, T.J.L.; Debije, M.G. Printed luminescent solar concentrators: Artistic renewable energy. *Energy Build.* **2020**, *207*, 27–30. [CrossRef]

110. Sethi, A.; Rafiee, M.; Chandra, S.; Ahmedj, H.; McCormack, S. A unified methodology for fabrication and quantification of gold nanorods, gold core silver shell nanocuboids and their polymer Nanocomposites. *Langmuir* **2019**, *35*. [CrossRef]

111. Aste, N.; Tagliabue, L.C.; Palladino, P.; Testa, D. Integration of a luminescent solar concentrator: Effects on daylight, correlated color temperature, illuminance level and color rendering index. *Sol. Energy* **2015**, *114*, 174–182. [CrossRef]

112. Aste, N.; Del Pero, C.; Tagliabue, L.C.; Leonforte, F.; Testa, D.; Fusco, R. Performance monitoring and building integration assessment of innovative LSC components. In Proceedings of the 5th International Conference on Clean Electrical Power: Renewable Energy Resources Impact, ICCEP 2015, Taormina, Italy, 16–18 June 2015; pp. 129–133. [CrossRef]

113. Fathi, M.; Abderrezek, M.; Djahli, F. Experimentations on luminescent glazing for solar electricity generation in buildings. *Opt. Int. J. Light Electron. Opt.* **2017**, *148*, 14–27. [CrossRef]

114. Vossen, F.M.; Aarts, M.P.J.; Debije, M.G. Visual performance of red luminescent solar concentrating windows in an office environment. *Energy Build.* **2016**, *113*, 123–132. [CrossRef]

115. Ahmed, H.; McCormack, S.J.; Doran, J. External quantum efficiency improvement with luminescent downshifting layers: Experimental and modelling. *Int. J. Spectrosc.* **2016**, *2016*, 1–7. [CrossRef]

116. Lamnatou, C.; Chemisana, D. Photovoltaic/Thermal (PVT) systems: A review with emphasis on environmental issues. *Renew. Energy* **2017**, *105*, 270–287. [CrossRef]

117. Shukla, A.; Kant, K.; Sharma, A.; Biwole, P.H. Cooling methodologies of photovoltaic module for enhancing electrical efficiency: A review. *Sol. Energy Mater. Sol. Cells* **2017**, *160*, 275–286. [CrossRef]

118. Phiraphat, S.; Prommas, R.; Puangsombut, W. Experimental study of natural convection in PV roof solar collector. *Int. Commun. Heat Mass Transf.* **2017**, *89*, 31–38. [CrossRef]

119. Hasanuzzaman, M.; Malek, A.B.M.A.; Islam, M.M.; Pandey, A.K.; Rahim, N.A. Global advancement of cooling technologies for PV systems: A review. *Sol. Energy* **2016**, *137*, 25–45. [CrossRef]

120. Kumar, A.; Baredar, P.; Qureshi, U. Historical and recent development of photovoltaic thermal (PVT) technologies. *Renew. Sustain. Energy Rev.* **2015**, *42*, 1428–1436. [CrossRef]

121. Calise, F.; Cappiello, F.L.; Dentice d'Accadia, M.; Vicidomini, M. Dynamic simulation, energy and economic comparison between BIPV and BIPVT collectors coupled with micro-wind turbines. *Energy* **2020**, *191*, 116439. [CrossRef]

122. Xu, L.; Luo, K.; Ji, J.; Yu, B.; Li, Z.; Huang, S. Study of a hybrid BIPV/T solar wall system. *Energy* **2020**, *193*, 116578. [CrossRef]

123. Debbarma, M.; Sudhakar, K.; Baredar, P. Comparison of BIPV and BIPVT: A review. *Resour. Technol.* **2016**, *3*, 263–271. [CrossRef]

124. Vats, K.; Tiwari, G.N. Performance evaluation of a building integrated semitransparent photovoltaic thermal system for roof and faade. *Energy Build.* **2012**, *45*, 211–218. [CrossRef]

125. Kamthania, D.; Nayak, S.; Tiwari, G.N. Performance evaluation of a hybrid photovoltaic thermal double pass facade for space heating. *Energy Build.* **2011**, *43*, 2274–2281. [CrossRef]

126. Agrawal, B.; Tiwari, G.N. Optimizing the energy and exergy of building integrated photovoltaic thermal (BIPVT) systems under cold climatic conditions. *Appl. Energy* **2010**, *87*, 417–426. [CrossRef]

127. Agathokleous, R.A.; Kalogirou, S.A. Double skin facades (DSF) and building integrated photovoltaics (BIPV): A review of configurations and heat transfer characteristics. *Renew. Energy* **2016**, *89*, 743–756. [CrossRef]

128. Yang, T.; Athienitis, A.K. A study of design options for a building integrated photovoltaic/thermal (BIPV/T) system with glazed air collector and multiple inlets. *Sol. Energy* **2014**, *104*, 82–92. [CrossRef]

129. Rajoria, C.S.; Agrawal, S.; Chandra, S.; Tiwari, G.N.; Chauhan, D.S. A Novel investigation of building integrated photovoltaic thermal (BiPVT) system: A comparative study. *Sol. Energy* **2016**, *131*, 107–118. [CrossRef]

130. Kim, J.H.; Park, S.H.; Kang, J.G.; Kim, J.T. Experimental performance of heating system with buildingintegrated PVT (BIPVT) collector. *Energy Procedia* **2014**, *48*, 1374–1384. [CrossRef]

131. Browne, M.C.; Norton, B.; McCormack, S.J. Phase change materials for photovoltaic thermal management. *Renew. Sustain. Energy Rev.* **2015**, *47*, 762–782. [CrossRef]

132. Baetens, R.; Jelle, B.P.; Gustavsen, A. Phase change materials for building applications: A state-of-the-art review. *Energy Build.* **2010**, *42*, 1361–1368. [CrossRef]

133. Chandel, S.S.; Agarwal, T. Review of cooling techniques using phase change materials for enhancing efficiency of photovoltaic power systems. *Renew. Sustain. Energy Rev.* **2017**, *73*, 1342–1351. [CrossRef]

134. Deo, A.; Mishra, G.K.; Tiwari, G.N. A thermal periodic theory and experimental validation of building integrated semi-transparent photovoltaic thermal (BiSPVT) system. *Sol. Energy* **2017**, *155*, 1021–1032. [CrossRef]

135. Tewari, P.; Mathur, S.; Mathur, J.; Kumar, S.; Loftness, V. Field study on indoor thermal comfort of office buildings using evaporative cooling in the composite climate of India. *Energy Build.* **2019**, *199*, 145–163. [CrossRef]

136. Ali, M.; Jamil, B.; Fakhruddin. Estimating diffuse solar radiation in India: Performance characterization of generalized single-input empirical models. *Urban Clim.* **2019**, *27*, 314–350. [CrossRef]

137. Makade, R.G.; Chakrabarti, S.; Jamil, B.; Sakhale, C.N. Estimation of global solar radiation for the tropical wet climatic region of India: A theory of experimentation approach. *Renew. Energy* **2020**, *146*, 2044–2059. [CrossRef]

138. Lam, N.L.; Smith, K.R.; Gauthier, A.; Bates, M.N. Kerosene: A review of household uses and their hazards in low-and middle-income countries. *J. Toxicol. Environ. Health Part B Crit. Rev.* **2012**, *15*, 396–432. [CrossRef]

139. Singh, Y.; Pal, N. Obstacles and comparative analysis in the advancement of photovoltaic power stations in India. *Sustain. Comput. Inform. Syst.* **2020**, *25*, 100372. [CrossRef]

140. Behuria, P. The politics of late late development in renewable energy sectors: Dependency and contradictory tensions in India's National Solar Mission. *World Dev.* **2020**, *126*, 104726. [CrossRef]

141. Kapoor, K.; Pandey, K.K.; Jain, A.K.; Nandan, A. Evolution of solar energy in India: A review. *Renew. Sustain. Energy Rev.* **2014**, *40*, 475–487. [CrossRef]

142. Raina, G.; Sinha, S. Outlook on the Indian scenario of solar energy strategies: Policies and challenges. *Energy Strateg. Rev.* **2019**, *24*, 331–341. [CrossRef]

143. Goel, M. Solar rooftop in India: Policies, challenges and outlook. *Green Energy Environ.* **2016**, *1*, 129–137. [CrossRef]

144. Yadav, S.K.; Bajpai, U. Energy, economic and environmental performance of a solar rooftop photovoltaic system in India. *Int. J. Sustain. Energy* **2020**, *39*, 51–66. [CrossRef]

145. Tarai, R.K.; Kale, P. Solar PV policy framework of Indian States: Overview, pitfalls, challenges, and improvements. *Renew. Energy Focus* **2018**, *26*, 46–57. [CrossRef]

146. Sethi, M. Climate change and urban areas. *Clim. Chang. Urban Settl.* **2018**, *86*, 1–44. [CrossRef]

147. Gupta, A.; Dalei, N.N. *Energy, Environment and Globalization: An Interface*; Springer: Singapore, 2020; ISBN 9789811393099.

148. Tulsyan, A.; Dhaka, S.; Mathur, J.; Yadav, J.V. Potential of energy savings through implementation of energy conservation building code in Jaipur city, India. *Energy Build.* **2013**, *58*, 123–130. [CrossRef]

149. Yu, S.; Tan, Q.; Evans, M.; Kyle, P.; Vu, L.; Patel, P.L. Improving building energy efficiency in India: State-Level analysis of building energy efficiency policies. *Energy Policy* **2017**, *110*, 331–341. [CrossRef]

150. Sharma, M. Development of a 'green building sustainability model' for green buildings in India. *J. Clean. Prod.* **2018**, *190*, 538–551. [CrossRef]

151. Chandel, S.S.; Sharma, A.; Marwaha, B.M. Review of energy efficiency initiatives and regulations for residential buildings in India. *Renew. Sustain. Energy Rev.* **2016**, *54*, 1443–1458. [CrossRef]

152. Ghosh, A.; Norton, B. Advances in switchable and highly insulating autonomous (self-powered) glazing systems for adaptive low energy buildings. *Renew. Energy* **2018**, *126*, 1003–1031. [CrossRef]

153. Ghosh, A.; Sarmah, N.; Sundaram, S.; Mallick, T.K. Numerical studies of thermal comfort for semi-transparent building integrated photovoltaic (BIPV)-vacuum glazing system. *Sol. Energy* **2019**, *190*, 608–616. [CrossRef]

154. Ravikumar, D.; Malghan, D. Material constraints for indigenous production of CdTe PV: Evidence from a Monte Carlo experiment using India's national solar mission benchmarks. *Renew. Sustain. Energy Rev.* **2013**, *25*, 393–403. [CrossRef]

155. Hairat, M.K.; Ghosh, S. 100 GW solar power in India by 2022—A critical review. *Renew. Sustain. Energy Rev.* **2017**, *73*, 1041–1050. [CrossRef]

156. Rathore, P.K.S.; Rathore, S.; Pratap Singh, R.; Agnihotri, S. Solar power utility sector in india: Challenges and opportunities. *Renew. Sustain. Energy Rev.* **2018**, *81*, 2703–2713. [CrossRef]

157. Tomar, V.; Tiwari, G.N. Techno-Economic evaluation of grid connected PV system for households with feed in tariff and time of day tariff regulation in New Delhi—A sustainable approach. *Renew. Sustain. Energy Rev.* **2017**, *70*, 822–835. [CrossRef]

158. Rohankar, N.; Jain, A.K.; Nangia, O.P.; Dwivedi, P. A study of existing solar power policy framework in India for viability of the solar projects perspective. *Renew. Sustain. Energy Rev.* **2016**, *56*, 510–518. [CrossRef]

159. Jain, S.; Jain, N.K. Cost of electricity banking under open-access arrangement: A case of solar electricity in India. *Renew. Energy* **2020**, *146*, 776–788. [CrossRef]

160. Heeter, J.; Vora, R.; Mathur, S.; Madrigal, P.; Chatterjee, S.K.; Shah, R. *Wheeling and Banking Strategies for Optimal Renewable Energy Deployment: International Experiences*; A Clean Energy Regulators Initiative Report NREL/TP-6A20-65660; United States Department of Energy (DOE): Washington, DC, USA, 2016; Volume 1, pp. 1–56.

161. Rathore, P.K.S.; Chauhan, D.S.; Singh, R.P. Decentralized solar rooftop photovoltaic in India: On the path of sustainable energy security. *Renew. Energy* **2019**, *131*, 297–307. [CrossRef]

162. Satsangi, K.P.; Das, D.B.; Babu, G.S.S.; Saxena, A.K. Real time performance of solar photovoltaic microgrid in India focusing on self-consumption in institutional buildings. *Energy Sustain. Dev.* **2019**, *52*, 40–51. [CrossRef]

163. Nundy, S.; Ghosh, A.; Mallick, T.K. Hydrophilic and superhydrophilic self-cleaning coatings by morphologically varying ZnO microstructures for photovoltaic and glazing applications. *ACS Omega* **2020**, *5*, 1033–1039. [CrossRef]

164. Chanchangi, Y.N.; Ghosh, A.; Sundaram, S.; Mallick, T.K. Dust and PV performance in Nigeria: A review. *Renew. Sustain. Energy Rev.* **2020**, *121*, 109704. [CrossRef]

165. Smestad, G.P.; Germer, T.A.; Alrashidi, H.; Fernández, E.F.; Dey, S.; Brahma, H.; Sarmah, N.; Ghosh, A.; Sellami, N. Modelling photovoltaic soiling losses through optical characterization. *Sci. Rep.* **2020**, 1–13. [CrossRef]

166. Kumar, A.; Prakash, O.; Dube, A. A review on progress of concentrated solar power in India. *Renew. Sustain. Energy Rev.* **2017**, *79*, 304–307. [CrossRef]

167. Bano, F.; Sehgal, V. Evaluation of energy-efficient design strategies: Comparison of the thermal performance of energy-efficient office buildings in composite climate, India. *Sol. Energy* **2018**, *176*, 506–519. [CrossRef]

168. Agrawal, B.; Tiwari, G.N. Life cycle cost assessment of building integrated photovoltaic thermal (BIPVT) systems. *Energy Build.* **2010**, *42*, 1472–1481. [CrossRef]

169. Gaur, A.; Tiwari, G.N.; Ménézo, C.; Al-Helal, I.M. Numerical and experimental studies on a building integrated semi-transparent photovoltaic thermal (BiSPVT) system: Model validation with a prototype test setup. *Energy Convers. Manag.* **2016**, *129*, 329–343. [CrossRef]

170. Karthick, A.; Kalidasa Murugavel, K.; Ghosh, A.; Sudhakar, K.; Ramanan, P. Investigation of a binary eutectic mixture of phase change material for building integrated photovoltaic (BIPV) system. *Sol. Energy Mater. Sol. Cells* **2020**, *207*, 110360. [CrossRef]

171. Karthick, A.; Murugavel, K.K.; Ramanan, P. Performance enhancement of a building-integrated photovoltaic module using phase change material. *Energy* **2018**, *142*, 803–812. [CrossRef]

172. Ghosh, A.; Norton, B. Optimization of PV powered SPD switchable glazing to minimise probability of loss of power supply. *Renew. Energy* **2019**, *131*, 993–1001. [CrossRef]

173. Nundy, S.; Ghosh, A. Thermal and visual comfort analysis of adaptive vacuum integrated switchable suspended particle device window for temperate climate. *Renew. Energy* **2019**, in press. [CrossRef]

174. Ghosh, A.; Norton, B.; Mallick, T.K. Influence of atmospheric clearness on PDLC switchable glazing transmission. *Energy Build.* **2018**, *172*, 257–264. [CrossRef]

175. Ghosh, A.; Mallick, T.K. Evaluation of colour properties due to switching behaviour of a PDLC glazing for adaptive building integration. *Renew. Energy* **2018**, *120*, 126–133. [CrossRef]

176. Ghosh, A.; Norton, B.; Mallick, T.K. Daylight characteristics of a polymer dispersed liquid crystal switchable glazing. *Sol. Energy Mater. Sol. Cells* **2018**, *174*, 572–576. [CrossRef]

177. Feng, Y.; Duan, Q.; Wang, J.; Baur, S. Approximation of building window properties using in situ measurements. *Build. Environ.* **2020**, *160*, 106590. [CrossRef]
178. Ghosh, A.; Norton, B. Interior colour rendering of daylight transmitted through a suspended particle device switchable glazing. *Sol. Energy Mater. Sol. Cells* **2017**, *163*, 218–223. [CrossRef]
179. Ghosh, A.; Norton, B.; Duffy, A. Effect of atmospheric transmittance on performance of adaptive SPD-vacuum switchable glazing. *Sol. Energy Mater. Sol. Cells* **2017**, *161*, 424–431. [CrossRef]
180. Ghosh, A.; Norton, B.; Duffy, A. Effect of sky clearness index on transmission of evacuated (vacuum) glazing. *Renew. Energy* **2017**, *105*, 160–166. [CrossRef]
181. Ghosh, A.; Norton, B.; Duffy, A. Effect of sky conditions on light transmission through a suspended particle device switchable glazing. *Sol. Energy Mater. Sol. Cells* **2017**, *160*, 134–140. [CrossRef]
182. Ghosh, A.; Norton, B. Durability of switching behaviour after outdoor exposure for a suspended particle device switchable glazing. *Sol. Energy Mater. Sol. Cells* **2017**, *163*, 178–184. [CrossRef]
183. Ghosh, A.; Mallick, T.K. Evaluation of optical properties and protection factors of a PDLC switchable glazing for low energy building integration. *Sol. Energy Mater. Sol. Cells* **2017**, *176*, 391–396. [CrossRef]
184. Ghosh, A.; Norton, B.; Duffy, A. First outdoor characterisation of a PV powered suspended particle device switchable glazing. *Sol. Energy Mater. Sol. Cells* **2016**, *157*, 1–9. [CrossRef]
185. Ghosh, A.; Norton, B.; Duffy, A. Daylighting performance and glare calculation of a suspended particle device switchable glazing. *Sol. Energy* **2016**, *132*, 114–128. [CrossRef]
186. Ghosh, A.; Norton, B.; Duffy, A. Measured overall heat transfer coefficient of a suspended particle device switchable glazing. *Appl. Energy* **2015**, *159*, 362–369. [CrossRef]
187. Ghosh, A.; Norton, B.; Duffy, A. Measured thermal & daylight performance of an evacuated glazing using an outdoor test cell. *Appl. Energy* **2016**, *177*, 196–203. [CrossRef]
188. Ghosh, A.; Norton, B.; Duffy, A. Measured thermal performance of a combined suspended particle switchable device evacuated glazing. *Appl. Energy* **2016**, *169*, 469–480. [CrossRef]
189. Hemaida, A.; Ghosh, A.; Sundaram, S.; Mallick, T.K. Evaluation of thermal performance for a smart switchable adaptive polymer dispersed liquid crystal (PDLC) glazing. *Sol. Energy* **2020**, *195*, 185–193. [CrossRef]
190. Michaux, G.; Greffet, R.; Salagnac, P.; Ridoret, J.B. Modelling of an airflow window and numerical investigation of its thermal performances by comparison to conventional double and triple-glazed windows. *Appl. Energy* **2019**, *242*, 27–45. [CrossRef]
191. Cuce, E. Accurate and reliable U-value assessment of argon-filled double glazed windows: A numerical and experimental investigation. *Energy Build.* **2018**, *171*, 100–106. [CrossRef]
192. Aguilar, J.O.; Xamán, J.; Olazo, Y.; Hernández-López, I.; Becerra, G.; Jaramillo, O.A. Thermal performance of a room with a double glazing window using glazing available in Mexican market. *Appl. Therm. Eng.* **2017**, *119*, 505–515. [CrossRef]
193. Li, D.; Li, Z.W.; Zheng, Y.M.; Liu, C.Y.; Lu, L.B. Optical performance of single and double glazing units in the wavelength 337–900 nm. *Sol. Energy* **2015**, *122*, 1091–1099. [CrossRef]
194. Wong, P.W.; Shimoda, Y.; Nonaka, M.; Inoue, M.; Mizuno, M. Semi-Transparent PV: Thermal performance, power generation, daylight modelling and energy saving potential in a residential application. *Renew. Energy* **2008**, *33*, 1024–1036. [CrossRef]
195. Lu, L.; Law, K.M. Overall energy performance of semi-transparent single-glazed photovoltaic (PV) window for a typical office in Hong Kong. *Renew. Energy* **2013**, *49*, 250–254. [CrossRef]
196. Ng, P.K.; Mithraratne, N.; Kua, H.W. Energy analysis of semi-transparent BIPV in Singapore buildings. *Energy Build.* **2013**, *66*, 274–281. [CrossRef]
197. Miyazaki, T.; Akisawa, A.; Kashiwagi, T. Energy savings of office buildings by the use of semi-transparent solar cells for windows. *Renew. Energy* **2005**, *30*, 281–304. [CrossRef]
198. Chow, T.T.; Pei, G.; Chan, L.S.; Lin, Z.; Fong, K.F. A comparative study of PV glazing performance in warm climate. *Indoor Built Environ.* **2009**, *18*, 32–40. [CrossRef]
199. Kiran Kumar, G.; Saboor, S.; Ashok Babu, T.P. Investigation of various wall and window glass material buildings in different climatic zones of India for energy efficient building construction. *Mater. Today Proc.* **2018**, *5*, 23224–23234. [CrossRef]
200. Gorantla, K.; Shaik, S.; Setty, A.B.T.P. Effect of different double glazing window combinations on heat gain in buildings for passive cooling in various climatic regions of India. *Mater. Today Proc.* **2017**, *4*, 1910–1916. [CrossRef]

Appl. Sci. **2020**, *10*, 2337

201. Skandalos, N.; Karamanis, D. PV glazing technologies. *Renew. Sustain. Energy Rev.* **2015**, *49*, 306–322. [CrossRef]
202. Saifullah, M.; Gwak, J.; Yun, J.H. Comprehensive review on material requirements, present status, and future prospects for building-integrated semitransparent photovoltaics (BISTPV). *J. Mater. Chem. A* **2016**, *4*, 8512–8540. [CrossRef]

MDPI

St. Alban-Anlage 66

4052 Basel

Switzerland

Tel. +41 61 683 77 34

Fax +41 61 302 89 18

www.mdpi.com

Applied Sciences Editorial Office

E-mail: applsci@mdpi.com

www.mdpi.com/journal/applsci